T0122204

Sustainable Civil Infrastructures

Editor-in-chief

Hany Farouk Shehata, Cairo, Egypt

Advisory Board

Khalid M. ElZahaby, Giza, Egypt
Dar Hao Chen, Austin, USA

Sustainable Infrastructure impacts our well-being and day-to-day lives. The infrastructures we are building today will shape our lives tomorrow. The complex and diverse nature of the impacts due to weather extremes on transportation and civil infrastructures can be seen in our roadways, bridges, and buildings. Extreme summer temperatures, droughts, flash floods, and rising numbers of freeze-thaw cycles pose challenges for civil infrastructure and can endanger public safety. We constantly hear how civil infrastructures need constant attention, preservation, and upgrading. Such improvements and developments would obviously benefit from our desired book series that provide sustainable engineering materials and designs. The economic impact is huge and much research has been conducted worldwide. The future holds many opportunities, not only for researchers in a given country, but also for the worldwide field engineers who apply and implement these technologies. We believe that no approach can succeed if it does not unite the efforts of various engineering disciplines from all over the world under one umbrella to offer a beacon of modern solutions to the global infrastructure. Experts from the various engineering disciplines around the globe will participate in this series, including: Geotechnical, Geological, Geoscience, Petroleum, Structural, Transportation, Bridge, Infrastructure, Energy, Architectural, Chemical and Materials, and other related Engineering disciplines.

More information about this series at http://www.springer.com/series/15140

Mona Badr · Ayman Lotfy

Editors

Sustainable Tunneling and Underground Use

Proceedings of the 2nd GeoMEast
International Congress and Exhibition
on Sustainable Civil Infrastructures,
Egypt 2018 – The Official International
Congress of the Soil-Structure Interaction
Group in Egypt (SSIGE)

 Springer

Editors
Mona Badr
German University in Cairo (GUC)
Cairo, Egypt

Ayman Lotfy
Ain-Shams University
Cairo, Egypt

ISSN 2366-3405   ISSN 2366-3413   (electronic)
Sustainable Civil Infrastructures
ISBN 978-3-030-01883-2   ISBN 978-3-030-01884-9   (eBook)
https://doi.org/10.1007/978-3-030-01884-9

Library of Congress Control Number: 2018957407

© Springer Nature Switzerland AG 2019
This work is subject to copyright. All rights are reserved by the Publisher, whether the whole or part of the material is concerned, specifically the rights of translation, reprinting, reuse of illustrations, recitation, broadcasting, reproduction on microfilms or in any other physical way, and transmission or information storage and retrieval, electronic adaptation, computer software, or by similar or dissimilar methodology now known or hereafter developed.
The use of general descriptive names, registered names, trademarks, service marks, etc. in this publication does not imply, even in the absence of a specific statement, that such names are exempt from the relevant protective laws and regulations and therefore free for general use.
The publisher, the authors and the editors are safe to assume that the advice and information in this book are believed to be true and accurate at the date of publication. Neither the publisher nor the authors or the editors give a warranty, express or implied, with respect to the material contained herein or for any errors or omissions that may have been made. The publisher remains neutral with regard to jurisdictional claims in published maps and institutional affiliations.

This Springer imprint is published by the registered company Springer Nature Switzerland AG
The registered company address is: Gewerbestrasse 11, 6330 Cham, Switzerland

Contents

An Approach for Optimization of Drilled Shaft Design in Dubai   1
Emad Y. Sharif and Hardev Sidhu

**Effect of Soil Parameter on the Long Drive Micro-tunelling Process a
Case Study of Al Jadaf Area, Dubai, UAE**   19
Devendra Datt Bhatt, Khalil Charif, and Rabee Rustum

**Probabilistic and Deterministic Analysis of an Excavation Supported
by Tiebacks and Nailing in Residual Soil of Gneiss**   31
Jean R. Garcia, Paulo J. R. Albuquerque, and Alexsander Silva Mucheti

**Ground Movements Induced by Tunnels Excavation Using
Pressurized Tunnel Boring Machine-Complete Three-Dimensional
Numerical Approach**   42
Mohammed Beghoul and Rafik Demagh

Convergence-Confinement Method Applied to Shallow Tunnels   53
Yassir Naouz, Latifa Ouadif, and Lahcen Bahi

**Poromechanical Behavior Analysis of an Underground Cavity Below
Runways Under the Dynamic Cyclic Action of Landing Gear on
Complex Geotechnical Conditions**   71
Mohamed Chikhaoui and Ammar Nechnech

**Earth Pressure Distribution on Rigid Pipes Overlain
by TDA Inclusion**   81
Mohamed A. Meguid

**Application of Electroosmosis and Monitoring for the Management
of Geotechnical Processes in Underground Construction**   94
Nicolay Perminov

**Ground Response and Support Measures for Rohtang Tunnel
in the Himalayas**   105
L. M. Singh, T. Singh, and K. S. Rao

**Engineering Failure Analysis and Design of Support System for
Ancient Egyptian Monuments in Valley of the Kings, Luxor, Egypt** . . . 126
Sayed Hemeda

**Recovery Process of the Water Distribution System After
a Seismic Event** 159
Yolanda Alberto-Hernandez

**Mechanically Stabilized Granular Layers - An Effective Solution
for Tunnel Projects** 169
Leoš Horníček and Zikmund Rakowski

In-Situ Evaluation of Adhesion Resistance in Anchor Root Zone 181
Burak Evirgen, Mustafa Tuncan, and Ahmet Tuncan

**Effect of Yield Criterion on Predicting the Arching Action Around
Existing Tunnels Subjected to Surface Loading** 191
Moataz Hesham Soliman, Sherif Adel Akl, and Mohamed Amer

**Failure Mechanism of Tunneling in Clay; Solution
by GFRP Reinforcement** 199
Zia Hoseini

Author Index 211

About the Editors

Ayman Lotfy Fayed, Ph.D., C.Eng., M.ASCE

– Professor (associate) of Geotechnical Engineering, Ain Shams University.
– Main research areas of interest: soil–structure interaction, deep foundations, excavation support systems, ground improvement, and subsurface characterization.
– Expert in data analysis and modeling using numerical approaches and artificial intelligence techniques.
– Extensive experience in teaching and training both undergraduates and graduated engineers.
– Geotechnical consultant of more than 20 years of professional practice.
– Broad experience in design and analysis of geotechnical systems including deep and shallow foundations, retaining structures, and ground improvement.
– Considerable experience in dredging and reclamation projects.

Education:

2002/Ain Shams University/Cairo, Egypt
Ph. D. in Civil (Geotechnical) Engineering
Thesis topic "Interaction between Deep Braced Excavation and Ground for Metro Subway Stations"

1997/Ain Shams University/Cairo, Egypt
M.Sc. in Civil (Geotechnical) Engineering
Thesis topic "Uplift Resistance of Shallow
Foundations"
1992/Ain Shams University/Cairo, Egypt
B.Sc. in Civil Engineering (Honor)

Associate Professor Dr. Mona Badr El-Din Anwar
is the instructor of the geotechnical engineering at the
German University in Cairo. She earned her B.Sc., M.Sc.,
and Ph.D. degrees from Helwan University, Egypt, in
1994, 1997, and 2001, respectively. She taught and
conducted research in most fields of geotechnical engineering as a university faculty member in Helwan
University (1995–2017) and in the German University
in Cairo (2014 till now). She is also a consultant
geotechnical engineer participating in planning, analyzing, designing, and providing follow-up for many
projects. She joined Dar Al-Handasah Consultants
(Shair and Partners) from 2008 to 2014; during this
period, she has been actively involved in many mega
projects in the Gulf area and in Africa. She is also member
of the Egyptian Code of Practice for Geotechnical
Engineering and Foundations on problematic soil committee, member of the ISSMFE, member of the ASCE,
and member of the ISRM.

An Approach for Optimization of Drilled Shaft Design in Dubai

Emad Y. Sharif[1](✉) and Hardev Sidhu[2]

[1] GTC Lab, Dubai, UAE
emad.sharif@gtc-lab.com
[2] Geotechnical Testing Expert, GTC Lab, Dubai, UAE
hardev.sidhu@gtc-lab.com

Abstract. Drilled shafts are widely used in Dubai as the main foundation option for high towers and bridges. Hundreds of meters of drilled shafts are installed in Dubai every day. Piling industry has grown widely in Dubai over the past 2 decades. Piles of large diameters as 2.5 m and lengths up to 100 m are common. It is evident that the current practice of geotechnical axial load capacity estimates in general is conservative, and actual pile axial load capacities as demonstrated by static and dynamic loading tests is much higher than the theoretical design capacities in almost the majority of cases.
The main purposes of this study is to review the shortcomings of the commonly used theoretical axial load estimate procedures, and examine the factors and parameters controlling the estimated axial capacity (input parameters). Optimization of the conservative design procedure is suggested by careful selection of main input parameters, and calibration to static loading tests.

Keywords: Drilled shaft · Rock mass factor · Rock compressive strength Load transfer

1 Very Breif Background on the Geology of Dubai

The rock formations noticed in Dubai area are mostly the mid to upper Tertiary sediments of Berzman Formation, consisting of an interbedded sequence of carbonate rich arenaceous, argillaceous and rudite/conglomeratic rocks. The weathered calcareous sandstones/Calcarenites with thin interbeds of siltstone are noticed to a depth of 18–25 m, underlain by the dominantly argillaceous/calci silttites with interbeds of polymictic conglomerates- with pebble to gravel size clasts of gabbros, chert, sandstones and carbonate rocks, Some deep-seated thick evaporate- gypsum beds associated with clays and mudstones are noticed beyond a depth of 80–100 m. These are continental shelf deposits with a depositional environment similar to the present day environment with frequent eustatic changes- fluctuations in the sea levels. The frequent sea level changes with harsh climatic conditions have given rise to the surfacial sabkah deposits during the quaternary times on wards along the coast lines.

© Springer Nature Switzerland AG 2019
M. Badr and A. Lotfy (Eds.): GeoMEast 2018, SUCI, pp. 1–18, 2019.
https://doi.org/10.1007/978-3-030-01884-9_1

2   Advantages of Drilled Shafts

The local geology in Dubai is characterized by extremely weak to weak (as per rock strength classification given in BS 5930 (1)) sedimentary rock of sandstone/siltstone/conglomerates/mudstone of compressive strength of 0.5–5.0 Mpa in general that is classified as an intermediate geo-materials (IGM) (2), or "Soft Rocks".

The use of drilled shafts in Dubai is attractive due to relative ease of drilling and advancing stable boreholes through the "soft" sedimentary rocks (IGMs) with time and cost effective drilling methods using augers/buckets. Bentonite muds and chemical polymers are typically used to support the drilled holes for piles of different diameters (as large as >2 m) and as deep as 100 m. Temporary steel casing is advanced through the upper un-consolidated sandy soil. In many projects, were drilled hole through IGM is stable, water only is used as drilling fluid, as adopted in many projects. Drilled shafts in IGM resist the applied loads by skin friction as opposed to end bearing. Design maximum side friction resistance ranging between 200 to 600 kPa are typical, depending on the local rock strength, mass parameters, and pile installation details.

3   Overview of the Behaviour of Typical Drilled Shafts Under Compressive Loading

A description of axial load resistance behavior of drilled shafts in IGMs is given in FHWA-RD-95-172 - Load Transfer for Drilled Shafts in Intermediate Geomaterials (3), which presents a parametric finite element study to assess and evaluate the different factors affecting the performance of drilled shafts and load transfer mechanisms. This study was calibrated by the results of actual in situ loading tests on instrumented piles. Some of important findings highlighted in this study are the determination of the many factors that control the magnitude and development/mobilization of the side friction and understanding load transfer mechanism. The IGM-Shaft interface conditions and the compressive strength of the IGM were found to be of primary significance. A rough interface would mobilize as large as 4 times the max side friction as a smooth interface, as shown on the graph (Fig. 1) extracted from this important study. The side friction was found to be highly affected by the existing initial horizontal effective pressure at the shaft – IGM interface, and hence the pile concrete fluidity (Slump). Therefore, the determination of compressive strength as uni-axial strength will under estimate the strength of IGM materials, and compressive strength is better to be estimated with tri-axial tests (CIU) at confining/consolidation pressure representing the in situ conditions.

FHWA-NHI-10-016 - Drilled Shafts Construction Procedures and LRFD Design Methods (4) also presents typical design procedure and performance prediction of drilled shafts based on actual behavior in static loading tests. The graph in Fig. 2 illustrates the load transfer behavior of a drilled shaft of length L and diameter B subjected to an axial compression load Q_T applied to the butt (top) of the shaft (Figure a). Figure b shows the general relationship between axial resistance and downward displacement, and figure c shows the load transfer along the shaft of the pile and end resistance mobilization. Several important behavioral aspects of drilled shafts are

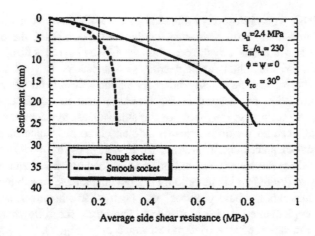

Fig. 1. Example of effect of interface roughness on load – settlement behavior infinite element analysis.

illustrated in Fig. 2. The first is that side and base resistances develop as a function of shaft displacement, and the peak values of each occur at different displacements. Maximum side resistance occurs at relatively small displacement and is independent of shaft diameter. Maximum base resistance occurs at relatively large displacement and is a function of shaft diameter and geomaterial type.

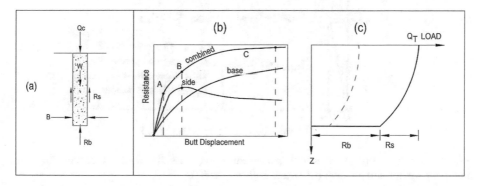

Fig. 2. Generalized load transfer behavior of drilled shaft in compression (FHWA-NHI-10-016).

4 Additional Design Considerations for Rock Sockets

A design decision to be addressed when using rock sockets is whether to neglect one or the other component of resistance (side or base) for the purpose of evaluating strength limit states. With regard to base resistance, AASHTO (5) Article C.10.8.3.5.4a states "Design based on side-wall shear alone should be considered for cases in which the base of the drilled hole cannot be cleaned and inspected or where it is determined that large movements of the shaft would be required to mobilize resistance in end bearing"

(AASHTO 2007). The philosophy embraced in the above comment gives a designer the option of neglecting base resistance. Under most conditions, the cost of quality control and assurance is offset by the economies achieved in socket design by including base resistance. Reasons cited for neglecting side resistance of rock sockets include (1) the possibility of strain-softening behavior of the sidewall interface, (2) the possibility of degradation of material at the borehole wall in argillaceous rocks, and (3) uncertainty regarding the roughness of the sidewall. Design for service limit states must, therefore, account for differences in side and base resistance mobilization as a function of axial displacement. Therefore, for drilled piles in Dubai, the side friction component provides the major contribution of axial load resistance and base resistance contribution is so limited and is ignored by many consultants. To enhance end bearing mobilization, methods as "base grouting" can be used (6), however, such techniques were not yet implemented in actual projects. Therefore, the following discussion is focused on optimization of side friction resistance only (Fig. 3).

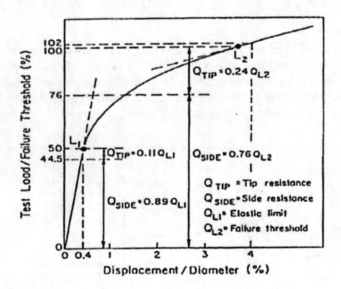

Fig. 3. Average normalized load-displacement curve that forms the basis of load test interpretation for compression (Chen and Kulhawy 2002).

5   Settlement Behaviour and Load Transfer Prediction

Simple empirical correlations for drilled piles exist that provide a simple, yet accurate estimates of load – settlement behavior (Fig. 4).

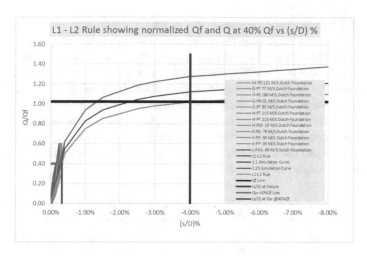

Fig. 4. Normalized pile load tests results plotted on L1–L2 graph.

5.1 L1–L2 Rule: Drilled Shaft Load - Settlement Behavior (Chen and Kulhawy 2002)

Data base of high quality static loading tests has been used to establish the commonly known L1–L2 rule (4); which describes the typical load – settlement and load transfer behavior of drilled shafts, as shown in the following schematic graph.

The L1–L2 normalized curve presents typical behavior of drilled shaft under axial load, and showing the transition from linear to non-linear and includes a definition of failure load (at 3–5% D settlement for pile socketed in IGM). The L1–L2 criteria was found to be applicable to results of static loading tests in Dubai and can be used to evaluate results of static loading tests, and as a simple design tool.

6 Local Drilled Pile Design Practice

The performance of drilled shafts in Dubai has been validated by many static loading tests on instrumented piles. In most cases, the results of static loading tests showed the actual capacity is much larger than the theoretical "nominal" design axial pile capacity. A study "Identification of soil and rock parameters by scientific interpretation of field measurements", 16th May, 2017 by Prof. Rolf Katzenbach was presented recently in Dubai showed a back calculation of results of static pile loading test showing that almost 50% of the pile length was only effective and necessary to resist the design load. The following graph shows results of many static pile loading tests conducted for piles within one major site in Dubai, plotted on the normalized L1–L2 graph, which indicates the actual capacity is much higher than the nominal design theoretical capacity.

The loading tests were conducted on working piles up to 150% of the nominal working load We. The nominal design is based on a global safety factor of 2.5, therefore, Qw = 0.4 Qf. By magnifying the L1–L2 simulation curve, it is found that the data fits with the 1.25 times magnified curve. Therefore, the actual pile design working load is >1.25 the adopted Qw, which leaves an important area for refinement of the theoretical design and make important savings. However, static loading tests conducted during construction stage on working piles are difficult to be used for design improvement. In few number of projects, tests are conducted on preliminary test piles during the design and pre-construction stage and has been effectively used to improve the theoretical design. Improvements of side friction resistance by >50% was achieved in several projects based on results of O-Cell pile loading tests during design stage. A typical example of 5 O-Cell tests conducted in one major project site showing maximum unit skin friction (fmax) versus theoretical estimated fmax from the site investigation results is shown in Fig. 5.

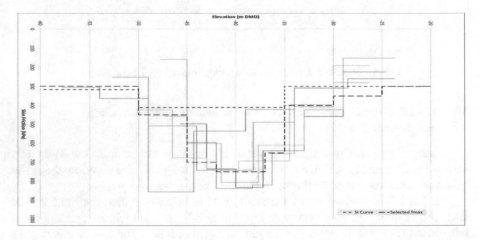

Fig. 5. O-Cell tests on PTP piles showing theoretical line (Red) and actual measured average unit friction.

The results of O-cell test for intervals far from the O-cell zone are under estimated because of limited development of friction resistance. Description of O-Cell tests & utilization of t-z curves for load test simulation and pile performance analysis are given in Fellenius (2014): "Basic of Foundation Design (7) & Unipile software (8). The following set of unit friction mobilization versus displacement (t-z curves) from one of the above O-cell test results is presented below, that shows limited unit friction is mobilized away from the cell zone (Fig. 6).

Therefore, optimization of the theoretical procedure for axial load estimate is of prime importance to bridge the gap between the conservative theoretical pile capacity estimates and actual pile behavior and achieve important savings. The use of preliminary tests piles is also encouraged as much as possible which provides reliable calibration of the theoretical assessed side friction and pile behavior.

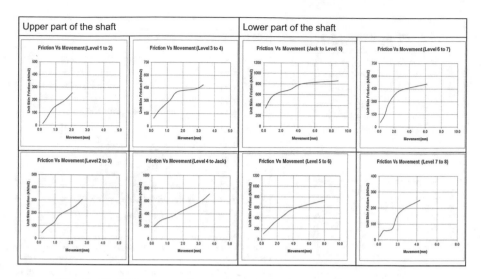

Fig. 6. Typical t-z curves for typical test pile (TP 8A – 1200 m diameter).

7 Theoretical Assessment of Maximum Side Friction in Local Practice

In local practice, the max. side friction (fmax) is typically estimated with some empirical equation as Horvath and Kenney (9), however, the procedure of Williams & Pells (10) is more common. Williams and Pells proposed method for skin friction evaluation is described in (10 & 11). The relationship between compressive strength of rock in socket and side-wall shear resistance, or adhesion factor is presented below (Fig. 7).

Bentonite Effect: The use of Bentonite slurry in advancing the pile shaft would cause the formation of bentonite cake causing smooth interface conditions. The use of other drilling fluids as chemical polymers would cause some improvement to skin friction. No data is available from local practice to provide an estimate of friction degradation due to method of pile installation. FHWA - NHI -10-016 (4), presents results of study (Brown 2002), in which side friction of piles advanced with different methods including dry polymer, liquid polymer, CAD (cased ahead), CFA (continuous Flight auger) and Bentonite, are compared.

In local practice, fmax is typically estimated assuming clean socket (where water or fluid polymer are used), however, where bentonite slurry is used, reductions of 20–30% are applied. The above Williams and Pells procedure provides a non-linear relationship (exponential) between the side friction adhesion factor and the rock strength. Non-linear relationships are more reliable than linear relationships as recommended by professional literature. According to Wiliams & Pells, procedure, the following main parameters determine the range of fmax:

a. Selection of representative Compressive Strength for each depth interval, and
b. Selection of reduction factor β,

For clean socket, the reduction factor β depends on (J) rock mass factor or modular ratio, as shown J-β relation (Williams & Pells) presented above.

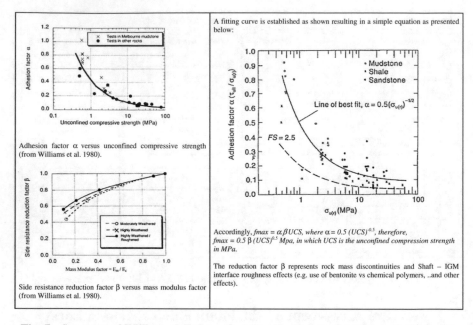

A fitting curve is established as shown resulting in a simple equation as presented below:

Adhesion factor α versus unconfined compressive strength (from Williams et al. 1980).

Side resistance reduction factor β versus mass modulus factor (from Williams et al. 1980).

Accordingly, $fmax = \alpha\beta UCS$, where $\alpha = 0.5\ (UCS)^{-0.5}$, therefore, $fmax = 0.5\ \beta\ (UCS)^{0.5}$ Mpa, in which UCS is the unconfined compression strength in MPa.

The reduction factor β represents rock mass discontinuities and Shaft – IGM interface roughness effects (e.g. use of bentonite vs chemical polymers, ..and other effects).

Fig. 7. Summary of Williams & Pells method for side friction of drilled shaft (10 & 11).

In local practice, J is estimated from RQD (Rock quality designation) using the relationship (Bieniawski 1984).

For most of the sites, RQD is <70% and hence a value of J < 0.2 is selected. Accordingly, the value of β < 0.65 is also selected (Fig. 8).

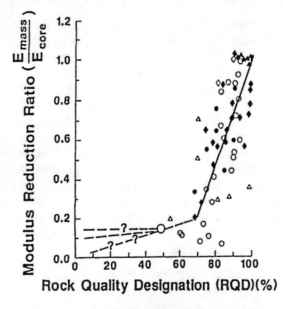

Fig. 8. Modulus reduction ratio as a function of RQD (From Bieniawski 1984).

Selection of design Compressive strength: The selection of compressive strength on the other hand is based on UCS (Uni-axial strength), which in the case of IGM provides a lower estimate of the rock strength. Further, large variations are typically obtained and therefore, the selection of design value for any depth interval becomes a difficult task! The graph presents typical UCS profile and selected design curve for a major project in Dubai (Fig. 9).

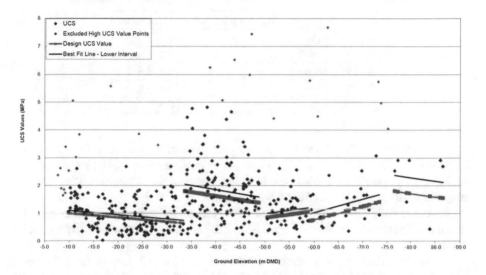

Fig. 9. Typical results of UCS tests profile showing large scatter of data, and showing selection of design UCS curve.

The graph shows that the profile is divided into "homogeneous" intervals, the extremely high UCS results are excluded, and design value which is less than the mean or best fit line depending on data variation and number of test data ("BS EN 1997-2004, Geotechnical Design, Part 1 – General Rules" and "A procedure for determining the characteristic value of a geotechnical parameter by A. J. Bond, Geocentrix Ltd., Banstead Surrey, United Kingdom".

The procedures for selecting both β and design UCS look satisfactory, however, the obtained fmax estimates are much less than actual resistance illustrated by actual loading tests.

Where preliminary pile loading tests (PTP) are conducted, a load test correction factor can be introduced as follows: fmax – corrected = $C_{correction}$ × fmax (from theoretical estimate).

However, for the general case, where PTP tests are not planned or considered, then the following optimization design steps are suggested.

8   Optimization of Design Procedure

Selection of reduction factor β: The mass factor J is related to drilling parameters as fracture index or Fracture Frequency per meter in addition to RQD as shown on the following Table 1 (Tomlinson: Pile Design & Construction Practice (12)).

Table 1. Mass factor J as function of RQD and fracture frequency/m.

RQD (%)	Fracture frequency per metre	Mass factor j
0–25	15	0.2
25–50	15–8	0.2
50–75	8–5	0.2–0.5
75–90	5–1	0.5–0.8
90–100	1	0.8–1

It is noticed that the above J – RQD relation is less conservative than the above Bieniawski, 1984 relationship. The following typical example shows results of high quality drilling results in a tower site in Dubai showing the different drilling parameters, and indicating RQD is <50–60% (Fig. 10).

Fracture Index vs Fracture Frequency: Fracture Index for all the cores (number of natural fractures/core run) – as measured for all BHs is shown in the following figure.

Figure 11 indicates a representative Fracture Index of <8, noting the core run is 1.5 m so the indicated results should be divided by 1.5 to five fracture frequency/m.

On the other hand J is defined as: E rock mass/E intact rock, and therefore it can be expressed as J = {(Vs-site)/(Vs-Lab)}^2, as Vs-lab represents shear wave velocity of intact core, and Vs-site represents rock mass quality. The ratio of Emass/Ecore may also represented by Er (re-loading modulus) measured with in situ pressure-meter tests (PMT) to CIU laboratory tests (E at 0.1% axial strain). PMTs and in situ shear wave velocity measurements are routinely conducted in Dubai but are not explicitly used for pile capacity design. For these tests results to be used to estimate the mass factor (J), CIU laboratory tests (consolidated undrained) tri-axial tests on intact core specimens of IGM with local strain and bender element tests shall be conducted. A typical result of multi-stage TX (CIU) test on IGM material is shown below presenting the local strain modulus vs axial strain. The results are very useful to estimate the representative modulus of IGM at given strain level, as 0.1% typically used for geotechnical analysis at service load level (Figs. 12 and 13).

The following Fig. 14 shows typical relationship obtained for the IGM modulus at 0.1 and 0.01% axial strain which represents small strain level.

Typical results from recently conducted investigation shows the correlation between PMT – Er and laboratory CIU modulus at 0.1% axial strain (local strain), which shows high ratio of about 1 between PMT – Er/Eu (local strain 0.1%), and indicating high mass factor J (Fig. 15.

The in situ measured shear wave velocity is presented in the following typical profile recently measured in a prime site in Dubai.

Fig. 10. Rock core parameters – all BHs.

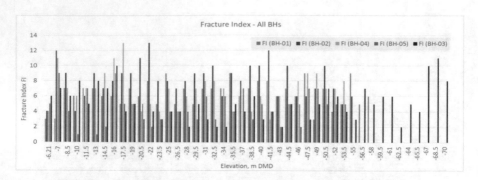

Fig. 11. Fracture index (fracture frequency) vs elevation – All Bhs.

The laboratory measured Vs using bender element tests in CIU tests (after consolidation) ranged between 600 to 1000 m/s, giving J (Mass factor) = (Vs-site/Vs-lab) ^2 of >0.4 in general for the sites included in this study.

Note: it is obvious that the global or total modulus typically measured and reported in standard site investigations from UCS tests is misleading and represents a fraction of the actual IGM stiffness. Measurements of global or total strain are affected by local stress concentration conditions near the ends of the test sample, and therefore, local strain is generally advised to be measured.

9   Compressive Strength of IGM

The uni-axial compressive strength is typically used to represent the strength of rocks. The compressive strength of Intermediate Geomaterials (IGMs), however, is highly affected by consolidation or confining pressure. High quality tests at GTC Lab in Dubai where recently conducted on many samples representing five main sites in Dubai showed a remarkable increase in compressive strength measured in CIU tests as consolidation pressure equal to in situ overburden pressure (taking Ko = 1). The tests were conducted using computer controlled tri-axial testing equipment, on high quality samples with local strain and bender element tests. Photos of UCS and TX test machines are shown in Fig. 16. All the tests were directly conducted by senior laboratory testing specialist (Hardev Sidhu).

Typical results are shown as follows: the CIU compressive strength is the deviatoric stress at failure. Linear trend line is added for each set of data for illustration only (Fig. 17).

The compressive strength (CS) obtained from CIU tests is at least 50% greater than UCS tests for different rock types (Calcarenite, Sandstone, Conglomerates, Siltstone and Calcisitite) (Fig. 18).

The above strength profile represents the following typical site lithology.

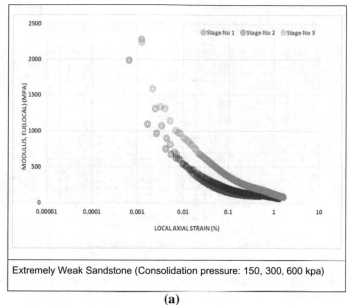

Extremely Weak Sandstone (Consolidation pressure: 150, 300, 600 kpa)

(a)

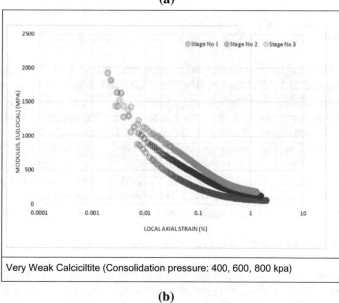

Very Weak Calciciltite (Consolidation pressure: 400, 600, 800 kpa)

(b)

Fig. 12. Modulus degradation obtained from instrumented CIU tri-axial test (Local Strain)

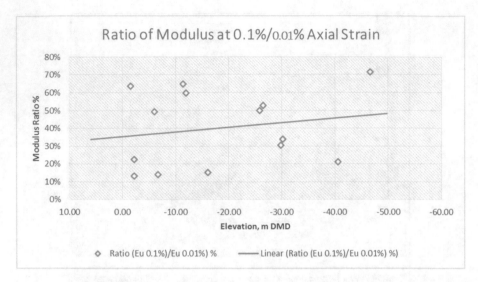

Fig. 13. Typical ratio of IGM modulus at 0.1–0.01% axial strain obtained from instrumented CIU Tri-axial tests.

10   Conclusion

Williams & Pells equation provides non-linear relationship with fmax, and includes parameters related to IGM (Intermediate Geo-Materials) -Shaft Interface (β), and rock strength. By combining the effects of proper selection of the mass factor J, and the use of Tri Axial based compressive strength, the following improvement is obtained:

Nominal maximum unit friction (fmax) = $0.5 \ \beta \ \sqrt{UCS}$

For typical β = 0.65 normally selected, then fmax = $0.5 \times 0.65 \ \sqrt{UCS}$ = $0.325 \ \sqrt{UCS}$

By proper selection of mass factor J, then β would range between 0.7 to 0.75, and by using triaxial compressive strength (CIU) test values rather than uniaxial/unconfined compressive strength (UCS) test values, with CIU shown to be 1.5 UCS in Dubai, then

$$\text{fmax} = 0.5 \times (0.7 - 0.75) \times \sqrt{(1.5 \times UCS)} \text{ or fmax} = (0.42 - 0.46) \times \sqrt{UCS}$$

An improvement factor of 1.3–1.4 over nominal estimates of fmax. This level of improvement is substantial and is generally matching with results of static loading tests.

It must be however, pointed out that the use of this range of parameters requires that:

High quality drilling to be conducted such that realistic rock mass parameters are obtained.

In situ PMTs and down-hole seismic tests to establish shear wave velocity and stiffness profiles representing the rock (IGM) mass.

Laboratory tri-axial (CIU) tests with consolidation pressure representing the overburden pressure, with local strain and few bender element tests.

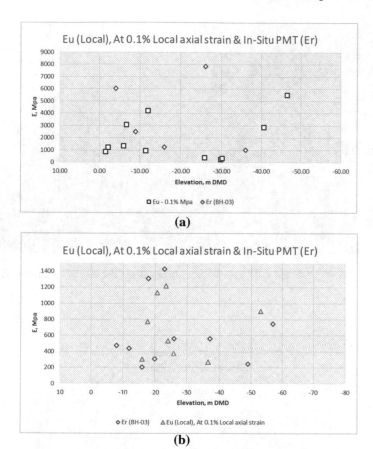

(a)

(b)

Fig. 14. (a) Laboratory modulus at 0.1% axial strain versus in situ pressure meter reloading modulus – site 1. (b) Laboratory Modulus at 0.1% axial strain versus in situ pressure meter reloading modulus – site 2.

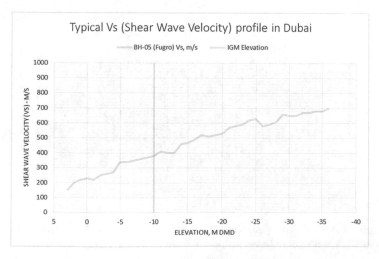

Fig. 15. Typical down hole seismic log of shear wave velocity profile in Dubai.

Fig. 16. Photos of UCS and CIU tri-axial computer controlled testing machines (GTC-LAB 2017)

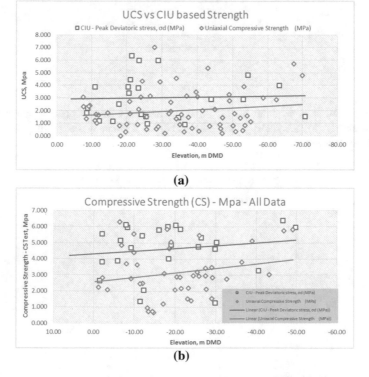

Fig. 17. (a) UCS vs CS (Compressive Strength) profile from CIU tri-axial tests – Site 1. (b) UCS vs CS (Compressive Strength) profile from CIU tri-axial tests – Site 2.

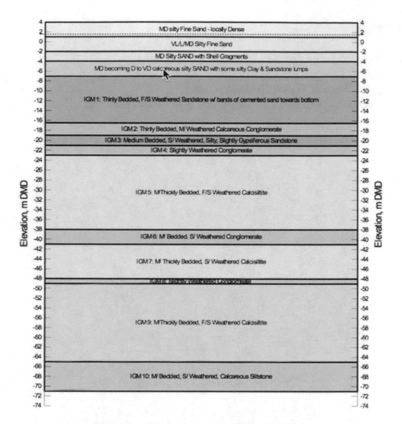

Fig. 18. Lithology of typical site in Dubai.

Most of the above tests are typically conducted in site investigations for main projects in Dubai, however, the use of TX – CIU testing was used only in few projects.

The number of tests and locations, and estimate of representative strength and stiffness parameters for each homogeneous depth interval shall follow standard geotechnical evaluation methods and must be conducted by experienced geotechnical engineer, to be able to establish appropriate estimate of mass factor J and hence friction reduction factor β, in addition to selecting representative design compressive strength (CS) values for different layers.

The above procedure makes the application of Williams & Pells formula much more reliable and less conservative and approach results of actual static loading tests.

References

BS 5930:2015 – Code of Practice for Ground Investigations
AASHTO LRFD Bridge Design Specifications, 4th edn. (2007)
Fellenius, B.H.: Basic of Foundation Design (2014)
Unisoft Geotechnical Solutions Ltd: Unipile 5.0 (Pile Design Software)

Effect of Soil Parameter on the Long Drive Micro-tunelling Process a Case Study of Al Jadaf Area, Dubai, UAE

Devendra Datt Bhatt[1](✉), Khalil Charif[2], and Rabee Rustum[3]

[1] Gulf Petrochemical Services and Trading L.L.C, Dubai, UAE
harshlovely@hotmail.com
[2] International Foundation Group L.L.C, Dubai, UAE
ifgllc@eim.ae, kcharif@ifguae.
[3] Water and Environmental Engineering, Heriot Watt University,
Dubai-Campus, Dubai, UAE
R.Rustum@hw.ac.uk

Abstract. This paper presents the effect of Soil Parameter on the long drive micro-tunneling process. Terzaghi's Silo Theory and Elastic Stress Relief Formulas are used for calculating the normal stress against the pipes. A detailed study tabulated for Al Jadaf Area across Oud Metha Road in Dubai, United Arab Emirates to check the effect of soil parameter, such as depth of excavation or invert level of micro-tunnelling, depth of ground water table, SPT Value and other soil properties. The study demonstrates that long term surface settlements confirm to the design theory limit. The high ground water table, fractures within the rock and close proximity to existing creek makes it very difficult to reduce the water table level in launching/receiving Pit and have difficulties in breaking the hard ground within the Pit. Determination of the surface settlement spot level difference is done physically at site by using latest surveying instrument, as described in the field exploration programme. By determining the difference between the two spot levels on a same coordinate, the amount of settlement and its effects on surface area can be established, which can be used as a guide line for future project and authentication of the theory.

Keywords: Overcut · Micro-tunnelling · Terzaghi's Silo theory
Ground water table · Jacking · Launch pit · SPT value

1 Introduction

There is a great amount of risk for surface and sub-structure settlement if soil parameters are not considered during the execution of micro-tunnelling. This is due to the fact that overcut is formed to provide space for converging ground and enables evenly distribution of surface lubrication fluid (bentonite) flow on the exterior pipes, as such, the buoyancy effects of fluid lubrication act on the pipe to lessen the drag during jacking.

Poor geological condition, low N value in the standard penetration test (SPT), measured in number of blow, can cause surface settlement, poor and unsuitable ground

© Springer Nature Switzerland AG 2019
M. Badr and A. Lotfy (Eds.): GeoMEast 2018, SUCI, pp. 19–30, 2019.
https://doi.org/10.1007/978-3-030-01884-9_2

tends to converge string of the pipe line. In hard rock, ground holes hold more stability and are conducive grounds to micro-tunnelling process [1]. The tendency of earth at forefront of micro-tunnelling to move viscously towards steel fitted in front of the pipe to be jacked should be considered. Therefore, sophisticated slurry shield, protection at the fore front of micro-tunnelling machine, are used to overcome this failure. Forces are relevant to face load depending upon the soil condition and its nature [2]. The front seal has the several advantages like safe working place for Men to enter and work; mounting for face stabilized device; a place to observe line and level and adjusting direction of the string.

Seals are control remotely and, in most cases, controlled by Computers. In Jadaf Area, where soil conditions are very poor and SPT values are very low and many services are close to receiving and launching Shaft. Therefore, a proper designing is required for dewatering process and special care is required to remove ground water only, and to ensure that soil particles, and another dissolved particles removal, are minimize to avoid any future surface settlement. Longer the length of micro-tunnelling greater will be frictional force, Resistance due to contact between soil and pipe, and thus can affect Jacking Load [3]. The man factors that can affect Jacking Load are disparity in ground condition; misalignment of pipes; roughness of pipe surface and joint; transient live load; Overcut; Interruption during Jacking; densification of loose sandy soil due to vibration/dynamic loading [4].

2　Field Exploration Programme

After obtaining the require NOC, No Objection Certificate, and other statuary requirements from relevant authority, spot level survey with the help of surveying auto level and co-ordinates by GPS[1] are taken at 2×2 m grid from 20 m to the either side from the centre of Micro-tunnelling between launch and receive pit prior to commissioning of the works. Similarly, 3D laser scanning method is used for micro-tunnelling reference no M01 & M04. After a minimum duration of one year from the date of completion similar spot level survey is conducted exactly at the same co-ordinate.

3　SPT Value and Its Effect on Actual Surface Settlement

Figure 1 shows the case study area for Jadaf, Dubai, United Arab Emirates with arrow mark near creek, which is recently connected to main sea via Dubai water canal project. In Fig. 2, complete pipe line network of 1200 mm diameter GRE, Glass Reinforced Epoxy, pipe with micro-tunnelling references highlighted in yellow. Complete network is divided in three stages. Soil parameter varies from M01 to M10 and accordingly the theoretical settlement and its comparisons with actual surface settlement is detailed in the study. The detail of the overall site plan is shown in Fig. 2. Light red colour lines

---

[1] Global Positing System.

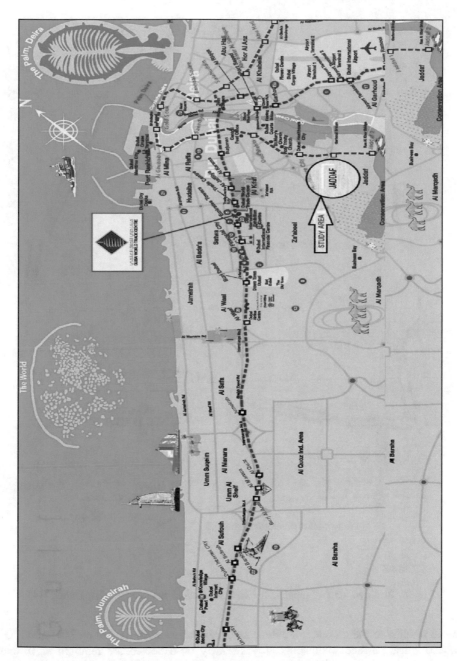

Fig. 1. Case study area of Jadaf, Dubai, UAE

are represented for pipeline and green for micro-tunnel. Source of the supply of the water is from south side after crossing the creek.

It is proved that the higher the N value the lesser the actual surface settlement (Please refer Fig. 7). Also, it is evident from Fig. 3 that greater SPT Value provide

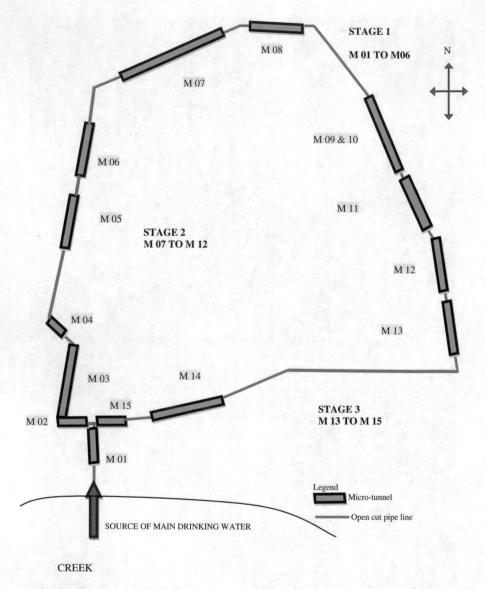

Fig. 2. Bird eye view of pipeline network with micro-tunnelling Jadaf, Dubai, UAE

more stability to the ground in all the cases, except for micro-tunnelling M01 and M04 where actual surface settlements are extremely un-expected are required to further scrutinize the method statement for spot level difference. Similarly, Table 1 gives the clear idea that higher the SPT value gives lower surface settlement. Higher the N Value means soil is dense and there is less chance for surface settlement [5, 6] (Table 2).

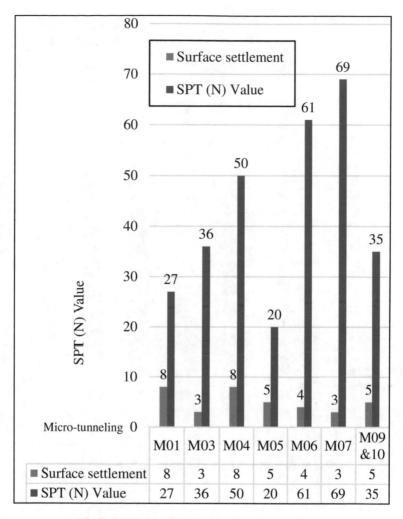

	M01	M03	M04	M05	M06	M07	M09 &10
Surface settlement	8	3	8	5	4	3	5
SPT (N) Value	27	36	50	20	61	69	35

Fig. 3. SPT value (N) Vs actual surface settlement

4 Effect of Ground Water Table on Actual Surface Settlement

Under the design theory, if water table is shallow then surface settlement is likely to be more and it is also reflected in the case study except for M01 and M04, where settlement is quite high (8 mm) with respect to their water level table. Reasons for this discrepancy might be the way spot level has been taken. Table 3 shows the maximum actual settlement and effect of ground water table and confirms to design theory [7, 8]. From Fig. 4, for micro-tunnelling M03 & 07, ground water table is 4 m and actual settlement observed at site by spot level method was found as 3 mm, whereas for others, if we calculate percentage between ground water table and surface settlement,

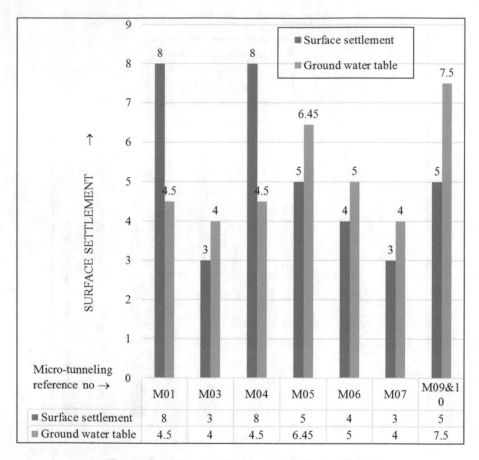

Fig. 4. Ground water table vs actual surface settlement

will find slight increase in the percentage as the water table depth increases. Though percentage difference very marginal such as for M03 it is 0.75, for M05 it is 0.78 and for M06 it is 0.80. Therefore, Ground water table has an impact on the micro-tunnelling process (Refer Fig. 8). It is also important to consider the dewatering process shall be in such a way so that silt/soil content remains within soil structure and does not pass through de-watering system. Also, slotted pipe or vertical screen pipe, generally PVC, and in special cases slotted steel, deep well must be of good quality and as per the design [9].

5   Local Conditions of Micro-Tunnelling in Jadaf Area

As the site area is in the city, the movement of heavy vehicles and noise is to be controlled accordingly. A special care is required for the summer time where cooling effect is necessary for the people working in the launching Pit. It is also necessary to control the temperature of air inside the tunnel; otherwise it can deflect laser control

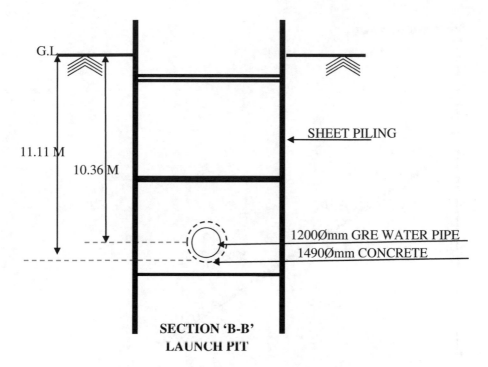

Fig. 5. Section of launch pit for micro-tunneling M09 & M10

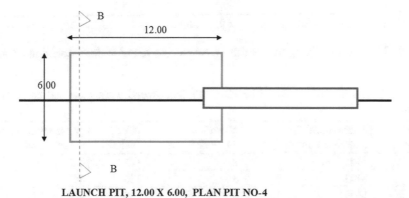

Fig. 6. Plan of launch pit for micro-tunneling M09 & M10

system and ultimately alignment of the tunnel. In Fig. 2 detail alignment of the pipeline and micro-tunnelling has been shown, to control overall movement of the traffic and other obstruction it is divided into three stages, all these are described in Table 3.

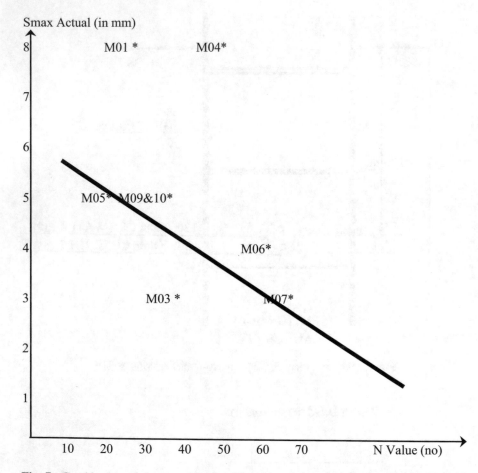

Fig. 7. Combined graph between N value (X-Axis) and Actual surface settlement (Y-Axis)

Table 1. Comparison of settlement between SPT (N), ground water table and actual surface settlement

Serial Number	Micro-tunneling reference number	Maximum Settlement mm	SPT value	Water table below ground level
1	M01	8	27	4.5
2	M03	3	36	4
3	M04	8	50	4.5
4	M05	5	20	6.45
5	M06	4	61	5
6	M07	3	69	4
7	M09&10	5	35	7.5

Table 2. Micro-tunnelling details for 1200 diameter GRE water transmission Pipe

S. n	Ref	Chainage		Length in meter	Stage	Line Ref No.	Location	Road reference
		From (L.P)	To (R.P)					
1	M-01	(-) 0 + 025.215	0 + 091.785	117	**Stage: 1 (796 M)**	18	Al Khail road crossing	E44
2	M-02	0 + 097.065	0 + 187.065	90		18	Existing services crossing (electric trough & 800 mm irrigation pipe)	Unmade area
3	M-03	0 + 493.865	0 + 192.865	301		18	Near Merriot hotel	Partially unmade area & one side service road crossing
4	M-04	0 + 499.510	0 + 637.510	138		18	Oud Metha road	E66
5	M-05	1 + 450.584	1 + 528.584	78		18	Service road crossing	Before EPCO petrol pump going towards Bur- Dubai
6	M-06	2 + 013.688	1 + 941.688	72		18	Service road crossing	After EPCO petrol pump going towards Bur -Dubai
7	M-07	(-) 0 + 009.943	0 + 304.734	309	**Stage: 2 (1,146 M)**	18A	Oud Metha road, WFI interchange	E66
8	M-08	0 + 397.734	0 + 304.734	93		18A	Service road crossing	Unmade area inside Latifa hospital
9	M-09	0 + 761.238	0 + 947.238	186		18A	Service road crossing	D79
10	M-10	1 + 137.313	0 + 951.813	186		18A		
11	M-11	1 + 146.205	1 + 404.205	258		18A	Existing services crossing	D79
12	M-12	1 + 524.468	1 + 410.468	114		18A	Existing services crossing (electric trough)	D79
13	M-13	0 + 458.688	0 + 686.688	228	**Stage - 3 (477 M)**	16A	Service road crossing	Near Jadaf Metro station
14	M-14	2 + 308.887	2 + 476.887	168		16	Service road crossing	Near Meriot hotel along Al Khail road towards Abu Dhabi
15	M-15	2 + 615.195	2 + 534.195	81		16	Service road crossing	Unmade area
Total Microtunneling Cummulative				**2419**				

6 Comparation on Different Soil Parameter, Theoretical and Actual Surface Settlement

Table 3 summarizes, soil parameter being used in this case study and overall summary of the result obtain from spot level difference with other details such as difference in the theoretical & actual surface settlement. It can be concluded that depth of micro-tunneling of the soil, ground water table, nature of soil, particle size distribution, diameter of the bore hole, has major effect on the surface settlement. For example, Fig. 5 is a typical cross-section at launching pit, where depth of invert level of the micro-tunnelling pipe is 11.10 m and depth of the axis of the pipe is 10.36 m, these are also reflected in Table 3, these data are required to work out surface settlement, which is 5.1 mm in this case. One of the bore log details is attracted herewith in Fig. 9. Figure 6 is the plan of the launching pit of M09 & 10, in which blue line represents sheet piling and red line is the centre of the micro-tunnelling longitudinally. Section B-B is shown in Fig. 5.

Table 3. Comparison of surface settlement between theoretical and actual with respect to soil parameter

S. nr	Micro-tunneling reference number	Maximum settlement (theoretical) mm	Maximum settlement (actual measure) mm	Length (m)	Depth of invert (m)	Depth of axis of pipe m	Water table below ground level (m)	SPT value (N)	Friction angle of soil (°)	Duration between spot level on same coordinate (day)
1	M01	4.4	8	117	12.67	11.93	4.5	27	40	338
2	M03	4.2	3	301	13.04	12.3	4	36	40	404
3	M04	4.4	8	138	12.64	11.9	4.5	50	40	400
4	M05	5.7	5	78	9.91	9.17	6.45	20	40	863
5	M06	6.4	4	72	9.01	8.27	5	61	40	639
6	M07	5.4	3	309	10.51	9.76	4	69	40	211
7	M09 & 10	5.1	5	372	11.1	10.36	7.5	35	40	515

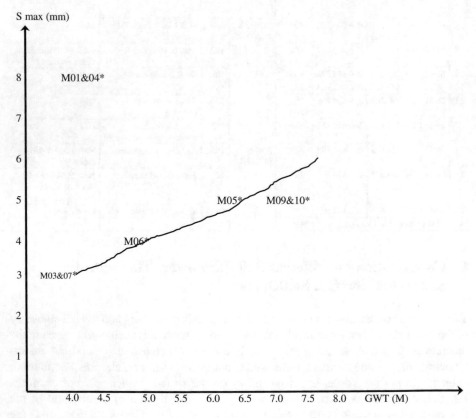

Fig. 8. Combined graph between Ground water Table (10-Axis) and Actual surface settlement (Y-Axis)

Location	Various Location in Dubai(Al Wasl Hospital)												
			Strata	Sample				Standard Penetration Test					
Depth	Reduced	Layer	Symbol	No	Test Depth		TCR	Number of blows				UCS	
(m)	Level (m)	Thick	W.L	&	(m)		[SCR]	Seating	Test Drive			N/mm2	Description of Strata
	[DMD]	(m)	Casing	Core			(RQD)	150	150	150	N Value		
			Depth	Run	From	To		mm	mm	mm			
0.5		0.5	X X	1 B	0.00	0.50		0	0	0			Brown, mottled grey slightly silty fine SAND with few gravel and shell fragments
1			X X	2 D	0.50	0.95		3	5	6	11		0.5m
1.5		1.5	X	3 D	1.00	1.45		3	6	8	14		Medium dense, brown slightly silty to silty medium to fine SAND with few cemented band and shell fragments
2			X X	4 D	1.50	1.95		4	7	10	17		2.0m
2.5		0.5	X X	5 D	2.00	2.45		2	4	6	10		Medium dense, brown, fine SAND with few cemented band and shell fragments
3			X X	6 D	2.50	2.95		7	10	15	25		2.5m
3.5			X	7 D	3.00	3.45		6	11	13	24		
4			X X										
4.5			X X	8 D	4.00	4.45		4	9	12	21		
5			X X										
5.5			X	9 D	5.00	5.45		6	14	19	33		Medium dense to dense, brown to light grey slightly silty to silty medium to fine/fine SAND with locally few cemented band and shell fragments
6		6.5	X X	10 D	6.00	6.45		5	11	16	27		
6.5			X										
7			X X	11 D	7.00	7.45		8	13	17	30		
7.5			▼ X										
8			X X	12 D	8.00	8.45		6	14	15	29		
8.5			X										9.0m
9			X X	13 D	9.00	9.45		8	14	18	32		Dense brown silty medium to fine SAND with few cemented bands and little gypsum
9.5		1.5	X										
10				14 D	10.00	10.45		10	17	18	35		10.5m E.O.B. 27.03.2011
10.5													
11													
11.5													
12													
12.5													
13													
13.5													
14													
14.5													
15													

Coordinates:	N : 2791062		E : 498768	Ground Level:		Ground Water Level:		7.5m	Date of Drilling:	27.03.2011
Drilling Method:	Cable tool percussion			Bore hole dia:	150mm	Casing Depth:		10.0m	Logged By:	
Core Barrel:	N.A			Core diameter:	N.A	Bit type:		N.A	Scale:	1:85
Remarks: Refer Plate No:		7	for Sample/ Test keys							

Fig. 9. Soil profile (Bore log for M09 & 10)

7 Overcut and Geological Condition

The case study has been conducted in a soil profile and geological condition (Please refer Fig. 9, soil profile), however other soil profile of other areas may represent different result. Similarly overcut, which is required to achieve workability and provide space for converging ground and evenly distribution of bentonite to be as per manufacturers recommendation, cutting wheel diameter of 1505 mm and a shield with external diameter of 1490 mm in front of shield that induce overcut equal to 7.50 mm each side of shield (difference between the external diameter and diameter of the outer surface of pipe string).

8   Conclusions

The present study has focused exclusively on the Jadaf area, Dubai, The United Arab Emirates, in a particular pattern of the soil with specific parameter for a given (1490 mm outer diameter composite pipe with concrete encashment, 1200 mm internal diameter, Glass Reinforce Epoxy pipe) pipe for main water transmission pipe line for DEWA. The bore and accordingly the micro-tunnelling machine has been selected for typical soil parameter, surrounded by existing underground services and various future proposal of infrastructure. Similar case study is required for other parameter of the soil structure and sub-soil geological condition such as detailed particle size distribution, rock strength, sulphite and chloride content etc. to ascertain challenges faced during construction and the amount of surface settlement within the range predicted in theory and after effect of the micro-tunneling. Therefore, we can conclude from this study that depth of micro-tunneling, ground water depth, nature of soil, particle size distribution and Diameter of Borehole has a major effect for surface settlement.

Acknowledgments. The authors would like to thank Mr Amer Ali Alsumaiti, NOC Manager, ROW, RTA Mr Ajaz Ahmed Abdul Rasheed, Chief Engineer RTA, and Dr Rigoberto Riverra, professor at UNAM for sharing with me their opinions.

References

1. Beard, A.N.: Tunnel safety, risk assessment and decision-making. Tunn. Undergr. Sp. Technol. **25**(1), 91–94 (2010)
2. Clarke, J.A., Laefer, D.F.: Evaluation of risk assessment procedures for buildings adjacent to tunnelling works. Tunn. Undergr. Sp. Technol. **40**, 333–342 (2014)
3. Bergeson, W.: Review of long drive microtunneling technology for use on large scale projects. Tunn. Undergr. Space Technol. **39**, 66–72 (2014)
4. Shi, P., Liu, W., Pan, J., Yu, C.: Experimental and analytical study of jacking load during microtunneling Gongbei tunnel pipe roof. J. Geotech. Eng. ASCE **144**(1), 1–8 (2018)
5. Barla, M., Camusso, M.: A method to design microtunnelling installations in randomly cemented Torino alluvial soil. Tunn. Undergr. Sp. Technol. **33**, 73–81 (2013)
6. Ong, D.E.L., Choo, C.S.: Back-analysis and finite element modeling of jacking forces in weathered rocks. Tunn. Undergr. Sp. Technol. **51**, 1–10 (2016)
7. ASTM, Standard Practice for Underground Installation of Thermoplastic Pressure Piping, ASTM, vol. i, no. Reapproved 2003, pp. 1–6 (2015)
8. Yen, J., Shou, K.: Numerical simulation for the estimation the jacking force of pipe jacking. Tunn. Undergr. Sp. Technol. **49**, 218–229 (2015)
9. Jebelli, J., Meguid, M.A., Sedghinejad, M.K.: Excavation failure during micro-tunneling in fine sands: a case study. Tunn. Undergr. Sp. Technol. **25**(6), 811–818 (2010)

Probabilistic and Deterministic Analysis of an Excavation Supported by Tiebacks and Nailing in Residual Soil of Gneiss

Jean R. Garcia[1]($\boxtimes$), Paulo J. R. Albuquerque[2],
and Alexsander Silva Mucheti[2]

[1] Federal University of Uberlândia (UFU), Uberlândia, Brazil
jean.garcia@ufu.br
[2] University of Campinas (Unicamp), Campinas, Brazil
pjra@fec.unicamp.br, alexmucheti@gmail.com

Abstract. With the growing economic appreciation of urban areas in large cities, building techniques that are technically and economically feasible have been widely employed to maximize utilization of spaces. Such techniques resort to large excavations to expand the land occupation rate. In this context, this paper analyzes a 24-m deep excavation located in São Bernardo do Campo, a city in the metropolitan region of São Paulo Metropolitan Region, Brazil. The local subsoil is composed by residual gneiss soil, and the retainment was executed in soil nailing. The length of the nails is variable in depth; the longer nails are introduced in the first 8 m of the excavation, and then reduced in greater depths. During the development of the retainment design, numerical analyses via limit equilibrium using deterministic and probabilistic methods considering the variations in properties of materials, demonstrated that the nails close to the end of the excavation are weakly influenced by the changes in nail lengths and diameters. In this sense, new analyses were performed to simulate the replacement of 1/3 of the nail rows for one or two rows of tiebacks at the end of the excavation to maintain stability of the retainment and reduce costs. The results for safety factor were obtained in deterministic and statistic (mean) terms. In the latter case, the reliability index and the probability of failure of the analyses performed are also presented. The results demonstrate the importance of conducting investigative analyses to determine the critical rupture surface by means of statistical analyses. In the case shown, this type of analysis proved to be more critical than the deterministic method of analysis of stability of retainment.

1 Introduction

The soil nailing technique has been widely used as a solution for retainment of large excavations in densely urbanized areas in several places worldwide, including Brazil. According to Liu, Shang, and Wu (2016), over the last decades soil nailing has become one of the most effective reinforcing methods, and it has been widely used in geotechnical structures such as excavation or slopes. In the first applications, soil nailing was considered as an alternative solution to conventional techniques to contain

© Springer Nature Switzerland AG 2019
M. Badr and A. Lotfy (Eds.): GeoMEast 2018, SUCI, pp. 31–41, 2019.
https://doi.org/10.1007/978-3-030-01884-9_3

excavations; it is now widely used in projects due to its acceptance and efficacy (Thomas 2003, 1997; Chassie 1993). According to Ghareh (2015), the advantage of soil nailing is its low cost, easy execution, consolidated technique and shorter time of execution, which has led to it being extensively used in engineering.

Technically, the nails prevent both soil strain and failure of the massif by transferring uplift forces generated by shear forces to the entire length of the inclusion (Luo, Tan, and Yong 2000). As a rule, the traditional method of safety coefficient is used to analyze stability of soil nailing. This method is based on the theory of limit equilibrium (Ugai and Leshchinsky 1995; Thomas 1997; Turner and Jensen 2005). However, retainment in nailed soil should be projected in order to be safety for all potential modes of rupture, including those caused by external and internal means and also on the wall surface. External ruptures refer to failures in global stability, sliding and support capacity at the bottom. In this type of analysis, the characteristics of the soil and of the nails are considered as homogeneous and all unknown, uncertain factors are attributed to a single significant coefficient.

According to Liu, Shang, and Wu (2016), there are uncertainties about these parameters: rupture may occur even if the calculated safety coefficient (FS) is greater that the minimum (FS_{min}) in conformance with the safety codes of the different countries. Therefore, it is important to carry out analyses on soil nailing stability that consider variability of the parameter of the soil and of the nails. In this sense, it becomes important to use ruin probability analysis to assess excavation stability (Babu and Singh 2009; Wright and Duncan 1991). Several studies have shown results on the variability of these parameters and the confidence intervals to be used in the analyses. According to Griffiths and Fenton (2004), the quick advances in IT have made it easier to use numerical methods that employ finite elements or finite differences methods in the retainment analyses.

In general, the analyses used routinely only check excavations in which the soil is considered as homogeneous, with permanent geotechnical parameters for a determined layer of soil. Therefore, it is necessary to better understand the behavior of retainments using the resource of probabilistic analyses instead of the analyses carried out conventionally via deterministic means. Therefore, this work has been conceived to analyze the safety of a retainment on nailed soil by means of ruin probabilistic analysis.

1.1 Safety Factors

The stability analyses of excavations can be performed via determination of two types of significant factors (FS), either via deterministic or probabilistic means. The deterministic safety factor is calculated for the minimum global sliding surface, from the stability analysis of regular slope (non-probabilistic). In this type of analysis, all input parameters are exactly like the mean values. In the probabilistic analysis, the mean safety factor is obtained, which is the mean value of all safety factors calculated for the minimum global sliding surface. In general, the mean safety factor should be close to the value of the deterministic safety factor. For a sufficiently large number of samples, the two values should be close; however, this does not take place in practice, since the samples analyzed have limited quantities and sometimes are not representative of the subsoil analyzed.

1.2   Variations in Soil Parameters

Significant variability is associated to soils and rocks in comparison to other engineering materials such as steel or concrete. Such variability increases the risk associated to forecasts (Ameratunga, Sivakugan, and Das 2016). Table 1 summarizes the values for coefficient of variation of some geotechnical parameters reported in the literature (Duncan 2000; Sivakugan, Arulrajah, and Bo 2011). In this sense, the variation coefficient is defined as:

Table 1. Typical values of the variation coefficient of some soil parameters (modified from Ameratunga, Sivakugan, and Das 2016)

Parameter	Coefficient of variation (%)
Unit weight, γ	3–7
Bouyant unit weight, γ'	0–10
Friction angle (sands), ϕ'	10
Friction angle (clays), ϕ'	10–50
Undrained shear strength, c_u	20–40
Standard penetration test blown count, N	20–40

$$CV(\%) = \frac{S}{\overline{X}} \cdot 100 \qquad (1)$$

Where: CV is the variation coefficient, S is the standard deviation and $\overline{X}$ is the mean.

For the analysis, the relative minimum and maximum values are converted into real minimum and maximum values when the statistic sampling is made for each random variable, as follows:

1.3   Types of Analysis

In this work, two types of analyses were conducted: the "minimum global analysis", which is performed via sampling method (Monte Carlo) from 1,000 samples. The minimum global is analyzed, i.e., the probabilistic analysis is conducted on the minimum global sliding surface located by the regular stability analysis (deterministic). This way, the safety factor is re-dimensioned N times (where N is the number of samples) for the "minimum global" sliding surface, using a different set of input variables randomly generated for each analysis.

1.4   Probability of Rupture and Reliability Index

The probability of failure is simply equal to the number of analyses with safety factor smaller than 1, divided by the total number of samples.

34   J. R. Garcia et al.

$$PF(\%) = \frac{N_{ruptura}}{N_{an\acute{a}lises}} \cdot 100 \tag{2}$$

Where: $N_{rupture}$ is the number of samples analyzed that produced a FS lower than 1, and $N_{analyses}$ is the total number of samples analyzed, in this case equal to 1,000.

The Reliability Index is another commonly used measurement of stability of slopes, after a probabilistic analysis is performed. The reliability index is an indication of the number of standard deviations that separate the mean safety factor from the critical safety factor (=1). The Reliability Factor can be calculated by assuming a normal distribution (Eq. 3) or a lognormal distribution (Eq. 4) of the results of the safety factor. The reliability index is usually equal to 3, which is the minimum recommended for safety of the stability project.

$$\beta_N = \frac{\mu_{FS} - 1}{\sigma_{FS}} \tag{3}$$

Where: β_N is the reliability index, μ_{FS} is the mean of the safety factor, and σ_{FS} is the standard deviation of the safety factor.

$$\beta_{LN} = \frac{\ln\left[\frac{\mu_{FS}}{\sqrt{1+V^2}}\right]}{\sqrt{\ln(1+V^2)}} \tag{4}$$

Where: β_{LN} is the reliability index, μ_{FS} is the mean of the safety factor, and V is the variation coefficient of the safety factor $(= \sigma/\mu)$.

1.5 Resistance of the Soil / Nail Contact

The friction strength between the soil and the nail was determined according to the experience in Brazilian soils, from back analysis of pullout tests (Fig. 1). The Eq. (5) below is based on the results of the simple sounding (N_{SPT}), per Eq. 3.

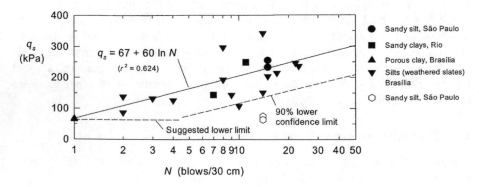

Fig. 1. Pullout tests results (Ortigão and Sayao 2004)

$$q_s = 67 + 60 \cdot \ln(N_{SPT}) \tag{5}$$

The allowable friction stress used in the design was determined for a safety factor equal to 2, per Eq. 6:

$$q_{adm} = \frac{q_s}{FS} \, (\text{FS} \geq 2) \tag{6}$$

1.6 Materials and Methods

The excavation area is located downtown São Bernardo do Campo city/SP, designed for construction of a15-floor apartment building with three underground floors. The excavation was executed including retainment of its sides in nailed soil, totaling 1,747 m² of sprayed concrete, 2,172 nails and 267 deep sub-horizontal drains (Cut 2017).

The nails were perforated with a 0.12 m diameter. The nails were previously assembled and then installed in the perforations. The nails included four sectors of injection besides the sheath (bore filling). The steel bars were of the CA-50 type. They were guarded with epoxy paint and the splices were made with steel sleeves pressed with a hydraulic device. The nails were filled and injected with cement slurry with a water/cement factor of 0.5 of weight. For the eight upper rows of nails, bars with a 20-mm diameter were used; in the eight center rows, the bars had a 25-mm diameter, and the eight lower lines had a 32-mm diameter.

The wall surface of sprayed concrete with thickness in the order of 0.10 m included an electro-welded metal mesh.

Wall drains (strips of geocomposite placed between the soil / sprayed concrete contact) and deep sub-horizontal split in six lines along the height of the retainment were used as the draining system.

The local subsoil is composed of gneiss residual soil and the retainment was executed on nailed soil. From the results of the investigations and subsoil characterization tests, the schematic profile was obtained with the key information on retainment, quantity and length of nails besides geological and geotechnical characteristics of the soil (Fig. 2).

1.7 Results and Analysis

Three potential failure surfaces were analyzed. These surfaces were established per recommendations from the geotechnical literature to check stability of nailed retainments and tieback retainments. Three potential failures surfaces were analyzed as determined per recommendations in the geotechnical literature to check stability of nailed and tieback retainments. (Figure 3). In the situations under analysis, the deterministic safety factor was determined by the Ordinary method (Fellenius).

After pre-defining the failure geometries (Fig. 3) and by means of the Slide 7.0® program (Rocscience), using the ruin analysis tool, it was possible to get the safety factors, probability of failure and the normal and lognormal reliability indexes. The

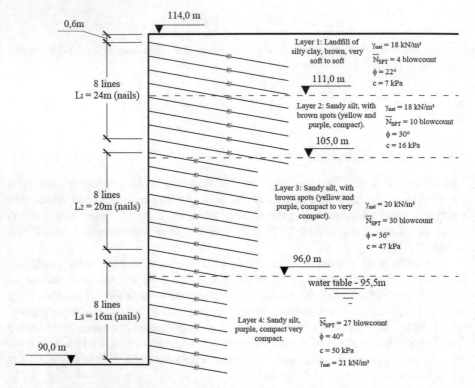

Fig. 2. Schematic profile of the retainment and soil properties

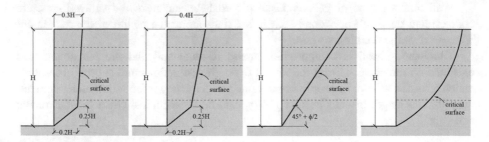

Fig. 3. Types of critical surfaces under analysis

analyses considered the failure surfaces indicated in Fig. 3 and several reinforcement conformations, as follows: 24 rows of nails (Figs. 4, 5, 6 and 13), 16 rows of nails and one row of tiebacks (Figs. 7, 8, 9 and 14), and 16 rows of nails and two rows of tiebacks (Figs. 10, 11, 12 and 15). For purposes of analysis, we checked the effect of decreasing the rows of nails from 24 to 16, and replacing for a tieback in one case, and by two tiebacks in another case.

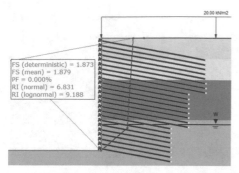

Fig. 4. Rupture surface in 0.3 H for 24 rows of nails

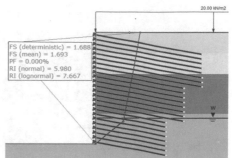

Fig. 5. Rupture surface in 0.4 H for 24 rows of nails

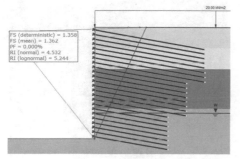

Fig. 6. Wedge-shaped rupture surface for 24 rows of nails

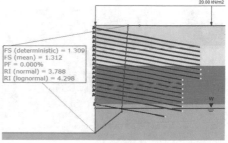

Fig. 7. Rupture surface in 0.3 H for 16 rows of nails and 1 row of tiebacks

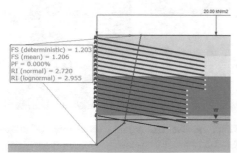

Fig. 8. Rupture surface in 0.4 H for 16 rows of nails and 1 row of tiebacks

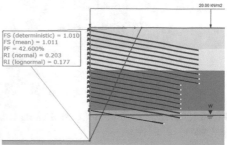

Fig. 9. Wedge-shaped rupture surface for 16 rows of nails and 1 row of tiebacks

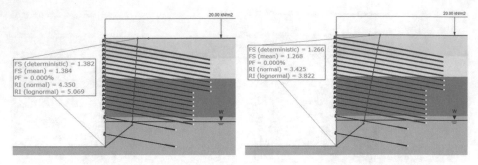

Fig. 10. Rupture surface in 0.3 H for 16 rows of nails and 2 rows of tiebacks.

Fig. 11. Rupture surface in 0.4 H for 16 rows of nails and 2 rows of tiebacks

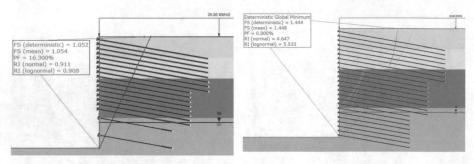

Fig. 12. Wedge-shaped rupture surface for 6 rows of nails and 2 rows of tiebacks.

Fig. 13. General rupture for 24 rows of nails

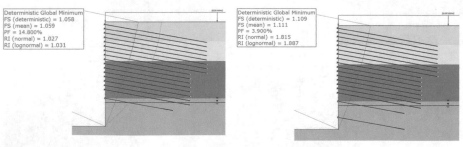

Fig. 14. General rupture for 16 rows of nails and 1 row of tiebacks

Fig. 15. General rupture for 6 rows of nails and 2 rows of tiebacks.

The results described in Table 3 show that, in all analyses made, the deterministic and mean (probabilistic) safety factors did not display significant differences, thus demonstrating that the number of samples analyzed can be considered appropriate.

The variability imposed to the soil parameters made it evident that the stability of the retainment was close to rupture (FS = 1) only for wedge-shaped rupture $(45° + \phi/2)$ (Figs. 9 and 12), with probability of failure at 42.6% and 16.3%, considering replacement of the last eight rows of nails for one and two rows of tiebacks respectively. However, in all analyses made for this case, the probability of failure was observed, since the reliability index was lower than the minimum ($\beta = 3$), which demonstrates the potential for failures. Por outro lado, as análises em termo de ruptura geral são satisfatórias (FS ≈ 1.45 e $\beta > 3$) somente para reforço integralmente efetuado com solo grampeado. For the option of replacing the last 6 nail lines by one and two tiebacks, the results of FS were 1.05 and 1.11. The probability of failure at 14.8% and 3.9%, respectively (Table 2).

Table 2. Soil resistance parameters

Layer	Interval	Young modulus	Cohesion	Friction angle	Specific weight
[-]	[m]	[kPa]	[kPa]	[°]	[kN/m^3]
1	114 –111	3,300	7	22	18
2	111–105	6,000	16	30	18
3	105–96	20,500	47	36	20
4	96–86	55,000	50	40	21

The values obtained for safety factors, ruin probability and reliability index for each situation can be seen in Table 3.

Table 3. Summary of the results obtained from the analyses

Surface	Reinforcement	FS$_{det}$	FS$_{mean}$	PF (%)	RI $_{(normal)}$	RI $_{(lognormal)}$
0.3 H	24 rows of nails	1.873	1.688	0.000	6.831	9.188
0.4 H		1.879	1.693	0.000	5.980	7.667
45° + φ/2		1.358	1.362	0.000	4.532	5.244
Global		1.444	1.448	0.000	4.647	5.533
0.3 H	16 rows of nails + 1 tieback	1.309	1.312	0.000	3.788	4.298
0.4 H		1.203	1.206	0.000	2.720	2.955
45° + φ/2		1.010	1.011	42.600	0.203	0.177
Global		1.058	1.059	14.800	1.027	1.031
0.3 H	16 rows of nails + 2 tiebacks	1.308	1.384	0.000	4.350	5.069
0.4 H		1.266	1.268	0.000	3.425	3.822
45° +φ/2		1.052	1.054	16.300	0.911	0.908
Global		1.109	1.111	3.900	1.815	1.887

Replacing eight rows of nails close to the end of the excavation (elevation 90 m) for one row of tiebacks significantly reduced the safety factor (FS $\cong$ 1). This can be interpreted as a probable risk to retainment stability. On the other hand, when a second row of tiebacks was added, in the case of the critical wedge-shaped surface, the observation was that the safety factor obtained remained unchanged and equal to 1.0, close to rupture.

An analysis of the rupture of the "face", for the distance of 0.3 H and 0.4 H upstream, showed that the option with two rows of tiebacks displayed a somewhat

1232

higher safety factor (Figs. 10 and 11) in comparison to the option of one row of tiebacks (Figs. 8 and 9), as an alternative to replacing the eight rows of nails close to the end of the excavation.

2 Conclusions

In general, deterministic analyses are used to analyze stability of retainment / excavations. In this case, the intrinsic variabilities of soil properties are not assessed. However, in some cases, they can point out factors of acceptable global safety (deterministic), but with values of ruin probability that indicate high risk of the work.

For the three situations of rupture reviewed in this paper, (0.3 H, 0.4 H and 45° + $\phi/2$), the conclusion was that the situations composed fully by nails (24 rows) displayed safety factors above 1.5 and a reliability index above 3.0. This demonstrates that, for this solution of retainment, the risk is either minimal or nonexistent.

In the case of replacement of eight rows of nails for one or two rows of tiebacks, the safety factor dropped to values below 1.5, to show a negative effect on the safety of the work. This fact was corroborated mainly via the reliability index below 3.0 in the cases of rupture surfaces 0.4 H and 45° + $\phi/2$. It must be pointed out that, in these cases, the probability of ruin is 42.6% and 16.3% for one and two rows of tiebacks respectively in the case of surface at 45°+ $\phi/2$. This indicates that, for the situation under analysis of 1,000 samples, 426 (one tieback) and 163 (two tiebacks) pose the risk of ruin. The analysis of the global failure was strongly influenced by the replacement of nail lines for tiebacks in the critical region near the end of the excavation.

The results obtained demonstrated the importance of conducting investigative analyses to determine the critical rupture surface by means of deterministic and probabilistic analyses. In the case under study, this type of analysis proved to be more critical than the deterministic method of analysis of stability of retainments since, besides the safety factor, values of probability of ruin and reliability index of the analyses can be obtained by means of variability of soil parameters.

Acknowledgments. The authors wish to thank UFU for purchasing the license of the software program used for the analyses in this paper.

References

Ameratunga, J., Sivakugan, N., Das, B.M.: Correlations of Soil and Rock Properties in Geotechnical Engineering (2016). https://doi.org/10.1007/978-81-322-2629-1
Babu, G.L.S., Singh, V.P.: Reliability analysis of soil nail walls. Georisk: Assess. Manag. Risk Eng. Syst. Geohazards **3** (1), 44–54 (2009).https://doi.org/10.1080/17499510802541425
Chassie, R.G.: Soil nailing overview and soil nail wall facing design and current developments. In: 10th Annual International Bridge Conference. Engineering Society of Western Pennsylvania, Pittsburgh (1993)
Duncan, J.M.: Factors of safety and reliability in geotechnical engineering. J. Geotech. Geoenvironmental Eng. **126**(4), 307–316 (2000). https://doi.org/10.1061/(ASCE)1090-0241 (2000)126:4(307)

Ghareh, S.: Parametric assessment of soil-nailing retaining structures in cohesive and cohesionless soils. Measurement, Elsevier Ltd. **73**, 341–351 (2015). https://doi.org/10.1016/j.measurement.2015.05.043

Griffiths, D.V., Fenton, G.A.: Probabilistic slope stability analysis by finite elements. J. Geotech. Geoenvironmental Eng. **130**(5), 507–518 (2004). https://doi.org/10.1061/(ASCE)1090-0241(2004)130:5(507)

Liu, J., Shang, K., Xing, W.: Stability analysis and performance of soil-nailing retaining system of excavation during construction period. J. Perform. Constr. Facil. **30**(1), C4014002 (2016). https://doi.org/10.1061/(ASCE)CF.1943-5509.0000640

Luo, S.Q., Tan, S.A., Yong, K.Y.: Pull-out resistance mechanism of a soil nail reinforcement in dilative soils. Soils Found. **40**(1), 47–56 (2000). https://doi.org/10.3208/sandf.40.47

Ortigão, J.A.R., Sayao, A.S.: Handbook of Slope Stabilisation. Springer-Verlag, Berlin Heidelberg GmbH, New York (2004)

Sivakugan, N., Arulrajah, A., Bo, M.W.: Laboratory Testing of Soils, Rocks and Aggregates. J. Ross Publishing, Fort Lauderdale (2011)

Thomas, J.T.: Stabilizing the stacks—a hybrid soil nailing system provides permanent underpinning for a historic college library. Civ. Eng. **67**, 44–46 (1997)

Thomas, J.T.: Soil nailing: where, when and why—a practical guide. In: 20th Central Pennsylvania Geotechnical Conference. Central PA Section of ASCE, Camp Hill (2003)

Turner, J.P., Jensen, W.G.: Landslide stabilization using soil nail and mechanically stabilized earth walls: case study. J. Geotech. Geoenvironmental Eng. **131**(2), 141–150 (2005). https://doi.org/10.1061/(ASCE)1090-0241(2005)131:2(141)

Ugai, K., Leshchinsky, D.: Three-dimensional limit equilibrium and finite element analyses: a comparison of results. Soils Found. **35**(4), 1–7 (1995). https://doi.org/10.3208/sandf.35.4_1

Wright, S.G., Duncan, G.M.: Limit equilibrium stability analyses for reinforced slopes. Transp. Res. Rec. **1330**, 40–46 (1991)

Ground Movements Induced by Tunnels Excavation Using Pressurized Tunnel Boring Machine-Complete Three-Dimensional Numerical Approach

Mohammed Beghoul[✉] and Rafik Demagh

University of Batna2, Batna, Algeria
mohammed.beghoul@mail.com

Abstract. The excavation of shallow tunnels in urban areas requires a previous evaluation of their effects on the existing constructions. In the case of shield tunneling, the different achieved operations is a highly three-dimensional soil/structure interaction. It is very complex to represent in a complete numerical simulation and therefore, the assessment of the tunneling-induced soil movements is difficult. In this paper, three dimensional simulation procedure, taking into account in an explicit and complete manner the main sources of movements in the soil mass, is proposed. It is illustrated in the case of Lyon's subway for which experimental data are available. The comparison of the numerical simulation results with the in situ measurements shows that the 3D procedure of simulation proposed is pertinent.

Keywords: 3D numerical simulation · Tunnel boring machine
Grouting · Confining pressure · In situ monitoring

1 Introduction

In urban areas, pressurized-face tunnel boring machines (TBMs), such as slurry shields, are increasingly employed to overcome the congestion at the ground surface. However, shield tunneling inevitably causes some ground movements that can have adverse effects on both buried and aboveground infrastructures and buildings. Therefore, it is a major concern in the underground works to estimate tunneling-induced ground movements due to shield tunneling. These ground movements are strongly influenced by many machine control parameters: excavation, front support, taper of the TBM shield, over-cutting, injection procedure and rheology of the grout. This complexity makes an explicit numerical simulation very difficult. In plane deformation, we can cite Pantet (1991), Moroto (1995), Kastner (1996), Benmebarek (2000). For these authors, the deformations are controlled using a single parameter, the deconfinement rate. In addition, empirical prediction such as numerical approaches that directly simulate the final phase may be insufficient. Several 3D numerical simulation procedures of tunnel excavation using pressurized-face tunnel boring machines, in soft and saturated soils,

© Springer Nature Switzerland AG 2019
M. Badr and A. Lotfy (Eds.): GeoMEast 2018, SUCI, pp. 42–52, 2019.
https://doi.org/10.1007/978-3-030-01884-9_4

have been proposed by various authors: Mroueh (1999), Dias (2000), Broere (2002), Kasper (2004), Bezuijen (2006), Demagh (2008, 2009a, 2009b), Lie (2015), Xie (2016). The confrontation with field results shows that, in spite of advances in terms of computing resources, the phenomena induced by the passage of a tunnel-boring machine are still insufficiently known.

This article presents a complete 3D numerical simulation of a tunnel boring process. This one is applied on the Lyon's subway project, for which well experimental data are available.

2 Ground Conditions and TBM

2.1 Overview of the Site

The project is part of the extension of the Lyon metro network, through an extension of the D line between the Gorge de Loup district and the Vaise station (Fig. 1). Two tunnels approximately 6.27 m in diameter and 900 m long were excavated successively between June 1993 and February 1995.

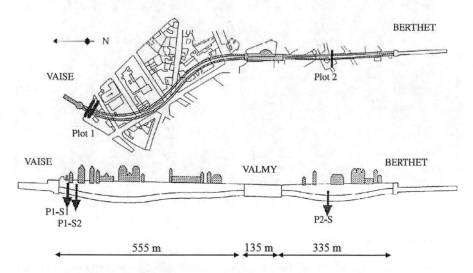

Fig. 1. Plan view and profile of the site

2.2 Experimental Support

The 3D numerical simulation is applied on the section P1-S1 of Lyon's subway for which experimental data are available (Table 1) This section is located 30 m from the starting point at Vaise, it is constituted of a superficial layer of fill, a layer of silty clay alluvium and a layer of sand and gravel (Fig. 2). The silt that is most often located at the crown of the tunnel has a natural water content very close to the liquid limit and a low consistency index, making it a very sensitive soil for reshuffling.

The presence of a retaining wall retaining an embankment six meters high makes the geometry of these two sections dissymmetrical.

Table 1. Mohr-Coulomb parameters

Layer	Depth m	G Mpa	K Mpa	γ kNm^{-3}	c' kPa	φ' degrees	ψ' degrees
Fill	0–7	2.92	7.8	18	30	30	17
Beige silt	7–10	2.74	7.3	19.5	12	33	20
Ocher silt	10–13	2.74	7.3	21	15	25	14
Grey clay	13–15	1.57	4.2	16.5	35	27	15
Grey sand	15–17.5	10.5	28	21	5	35	21
Purplish clay	17.5–20	5.16	13.8	18.5	35	27	15
Sand and gravel	20–30	10.5	28	21	0	34	20

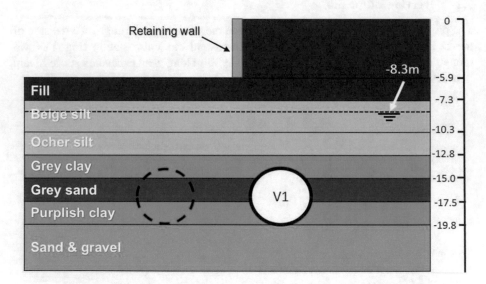

Fig. 2. Geological profile

2.3 Slurry Shield TBM

The knowledge of the nature of the terrain, the route located largely under the built-up area (old masonry buildings) and the low coverage led the group of companies to propose the technique of the slurry shield TBM. This tunnel-boring machine consists of a shield (slightly conical) of about 7 m long and 6.27 m in diameter (Table 2), a 50-meter long back-up train and weighing nearly 300 tons. The machine is equipped with about fifty sensors recording the parameters of continuous operation. The tunnel is covered with prefabricated concrete segments (35 cm thick) placed under the shelter of the shield tail: when the tunnel boring machine moves forward and the ring escapes from the shield tail, there remains an annular void of 13.5 cm (thickness of the shield tail) between the soil and the lining (Fig. 3). In order to minimize settling, this void is immediately injected with a filling grout made of bentonite, sand and water.

Table 2. Tunnel parameters

Section	H m	L m	C/D -	Hw m	Dext m	Dint m	Δ/2 mm	Φext m	Φint m	Liner width cm	Liner thick cm
P1-S1	16.9	7.0	2.20	8.3	6.27	6.24	15	6.0	5.3	100	35

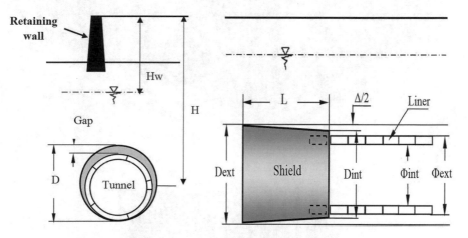

Fig. 3. Tunnel, shield and liner parameters

3 Computational Modelling

3.1 Finite Difference Model

To assess the ground movements during slurry shield tunneling, a finite difference (FD) model is developed using Itasca Flac3D. The FD model is developed (Fig. 4) with zero transverse (x-axis) displacement at $x = -120$ m and $x = 120$ m, and zero longitudinal (y-axis) displacement at $y = 0$ and $y = 120$ m, respectively. The top of the model boundary ($z = 0$) is set to be free, whereas the vertical movement at the bottom boundary ($Z = -30$ m) is fixed. For this model, average soil parameters determined from the geotechnical tests are used (Table 1). The soil is modelled using solid elements with 8-grid points (155000 grid points). A linear elastic, perfectly plastic Mohr Coulomb constitutive model is employed with no-associative flow rule. The retaining wall is modelled by an elastic rule. The extent of the model, in the longitudinal direction, is conditioned by the position of the stationary section (Fig. 4). The model element lengths equal the ring width of 1 m in the longitudinal direction within the tunnel excavation zone.

A shield with conical shape, perfectly rigid (the nodes are fixed according to the method called fixed center) Benmebarek (2000), modelled with thin volumetric elements, is installed in a virgin ground solid mass (Fig. 5) for which an initial state of geostatic stresses is imposed with K0 equal to 0.5 (normally consolidated soils), Benmebarek (2000), Demagh (2008, 2009a, 2009b).

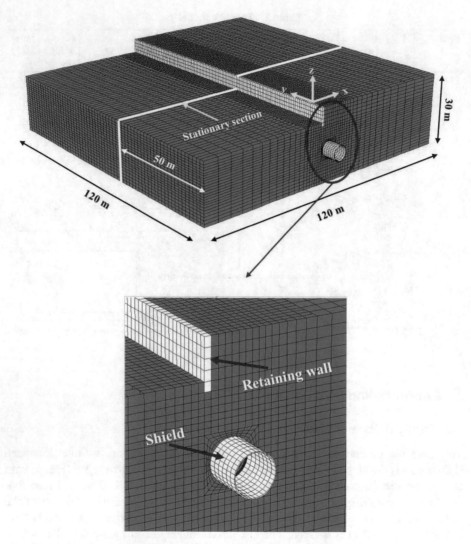

Fig. 4. Flac3D finite difference model of slurry shield TBM tunneling.

The geometry of this section made the three-dimensional simulations more difficult, due to the presence of a retaining wall on the surface. This singularity forced us to consider the three-dimensional model in its entirety. On the other hand, the three-dimensional numerical models found in the literature preferred vertically symmetric models, which made the elaborate simulation procedures less constraining.

From the moment, that the shield is completely installed (Fig. 5), the modelling sequence (Fig. 6a) could be applied.

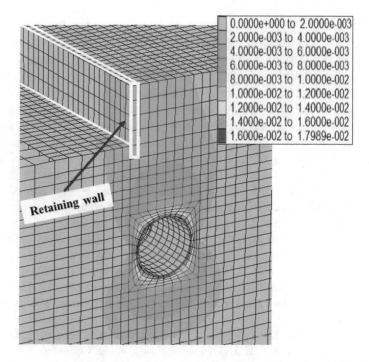

Fig. 5. Contour of vertical displacements after complete installation of the shield (units: m)

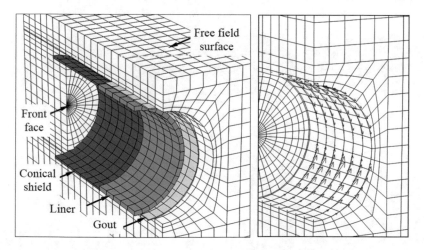

Fig. 6. Used mesh (a) View of different parts (b) Arch effect of the displacements at the soil/shield interface

3.2 Modelling Sequence

The excavation is simulated by the deactivation of disk element equal to the length of the lining segment (1 m). The stability of the front face is not the most important source of surface settlement Benmebarek (2000), so in our simulations, we considered that displacements at the front of the excavation are blocked by the shield. The shield passage, simulated by annulment of local tangential stresses, clears a volume loss that is immediately filled by soil convergence (very poor soils around the excavation). The interface which is on the shield is activated from the moment since a contact is established with the surrounding ground: the role of this interface is to block the radial ground convergence and also to allow the tangential convergence by arch effect (Fig. 6b).

The volume loss is partially compensated by the possible migration of the grout forwards the face of the shield (there is a great uncertainty on the post-closing shape of the ground around the shield). This migration is simulated by the correction of the shield conical shape, set so as to reproduce a vertical displacement recorded on construction site (back analysis on surface and/or tunnel crown vertical displacement), Dias (2000) and Demagh (2008). The liner is simulated by shell elements. It is characterized by a weaker Young modulus in order to take account of the seals between the prefabricated rings of liner. The injection of the grout in the annular void is controlled in volume and pressure. The position of the injection pipes was decisive in the choice of the pressure diagram. The maximum value of the injection pressure is fixed on the measure of a maximum vertical displacement closest to the vertical axis of the tunnel (recorded on building site). It shows in particular that the pressure really transmitted to the ground remains lower than the average pressure measured at the exit of the injection pipes. This difference is due to the pressure loss by friction following the flow of the

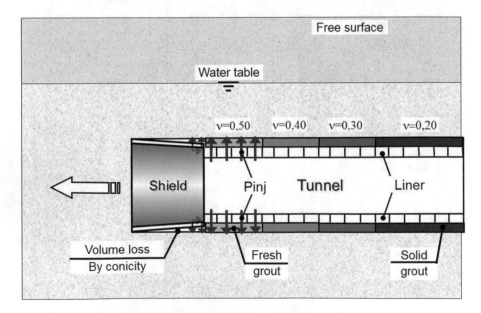

Fig. 7. Complete phased simulation of TBM excavation process

grout like to its impregnation of the surrounding ground Demagh et al. (2008). Uncertainty on the rheology of the grout brings to consider two principal phases (liquid and solid phase) intercalated by a transitional one (Fig. 7).

This procedure is repeated throughout the shield progression, until reaching a stationary section (Fig. 4) after a few tens of excavation steps. The simulations are carried out in drained conditions (effective stresses with taking into account of the submerged volumetric weight corresponding to the long-term behaviour). The simulation results are confronted with in situ data collected on construction tunnel site.

4   Results

Figure 8, represents the confrontation of the in situ trough with the 3D simulation results. We note a good agreement in shape and amplitude between the two graphs. This can be explained by the good simulation of the different phases of the tunnel boring process. The effect of the retaining wall is also well represented by the 3D simulations. The amplitude of the trough (11 mm) can be explained by the fact that this section located at a break-in area where the operating parameters of the tunnel-boring machine are in a test period.

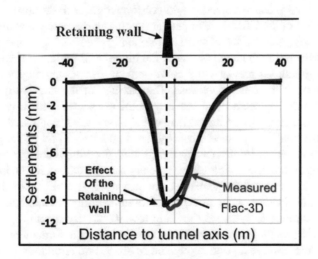

Fig. 8. Transverse settlement trough at final state

Figures 9, illustrates the ability of the 3D procedure to describe the effect of the TBM progression through the longitudinal settlement trough, what shows the relevance of the choices for the simulation, in particular the maximum value of the injection pressure, fixed on the vertical displacement of depth point on central extensometer recorded on building site. In addition, the choice of the injection pressure distribution is justified by the grout ports position at the back of the shield tail.

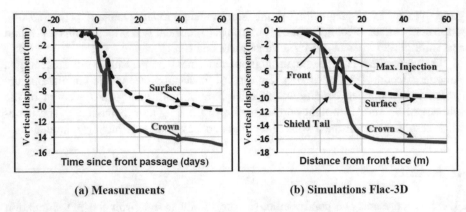

(a) Measurements   (b) Simulations Flac-3D

Fig. 9. Evolution of longitudinal settlements (surface and crown)

In our simulations, the progress of the shield (x-axis) is measured by distance from the front (the shield excavates one meter a time in our simulations) (Fig. 9b). We couldn't measure it in days because the progression of the TBM in situ is not constant.

The progression of the TBM could be summarized in four principal phases (front passage, shield tail passage, injection and consolidation of the grout):

The influence of the TBM is felt upstream with approximately one diameter before the measurements section by a small settlement. The front passage results in an equivalent settlement in the key point near the tunnel axis and on surface. At the front passage, the rigid shield prevents any movement of ground convergence; the excavation is thus done with constant volume. The passage of the tail shield results by an additional vertical displacement of the key point associated with a weak additional surface settlement (Fig. 9). These settlements are very weak compared to the formed gap behind: this can be explained either by the migration of part of the grout injected under pressure forwards by decreasing the ground decompression, or with the insufficiency of time so that the soil closes completely the annular void. The injection of the grout heaves the crown of the tunnel of about 5 mm. The last phase is the differed settlements (consolidation of the grout) which is started after the maximum effect of the injection and continue to evolve for tending towards an asymptote.

Figure 10, shows the lateral movements (inclinometric deformations) on both sides of the tunnel. We observe a good agreement in shape and amplitude between the two graphs especially around the tunnel axis. However, in the upper part of the tunnel, the in situ results are not well reproduced by the 3D simulations. This can be explained by the many parameters involved in the 3D simulation procedure, but also by the use of the Mohr-Coulomb model which is not adapted to this type of soils.

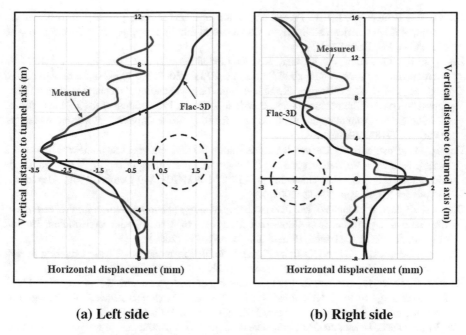

(a) Left side   (b) Right side

Fig. 10. Evolution of inclinometric deformations at final state

5 Conclusion

A 3D simulation procedure has been proposed to account for all the different operations achieved by a TBM. This procedure has been applied to reproduce by back-analysis the movements recorded on the P1-S1 section of Lyon's subway.

The comparison of the results of simulations with the field measurements, shows that the proposed 3D simulation procedure is pertinent, in particular in the representation of the different operations carried out by the tunnel boring machine.

Singularity such as the presence of the retaining wall is well simulated.

However, uncertainties related to grout injection remain; if the migration of the grout seems to be well simulated by a corrected conicity, the different phases of the injection are still difficult to simulate.

Finally, If we want to simulate the impact of the main excavating parameters, only a 3D simulation procedure taking into account each of them would in principle make it possible.

References

Benmebarek, S., Kastner, R.: Modélisation numérique des mouvements de terrain meuble induits par un tunnelier. Revue Canadienne de Géotechnique **37**, 1309–1324 (2000). (in french)

Bezuijen, A., Talmon, A.M.: Bakker, et al. (eds.) Proceedings of the Geotechnical Aspects of Underground Construction in soft Ground, pp. 187–193. Taylor & Francis Group, London (2006)

Broere, W., Brinkgreve, R.B.J.: Phased simulation of a tunnel boring process in soft soil. In: Mestat, (ed.) Numerical Methods in Geotechnical Engineering, pp. 529–536. Presses de l'ENPC/LCPC, Paris (2002)

Demagh, R., Emeriault, F., Kastner, R.: 3D modelling of tunnel excavation by TBM in overconsolidated soils. JNGG 2008 Nantes, 18–20 Juin 2008, pp. 305–312 (in french) (2008)

Demagh, R., Emeriault, E., Kastner, R.: Shield tunnelling -validation of a complete 3D numerical simulation on 3 different case studies. Euro:Tun 2009. In: Proceedings of the 2nd International Conference on Computational Methods in Tunnelling. Ruhr University Bochum, September 2009, pp. 77–82 (2009a)

Demagh, R., Emeriault, F., Kastner, R.: Modélisation 3D du creusement de tunnel par tunnelier à front pressurisé – Validation sur 3 cas d'études. In: Proceedings of the 17ème Conférence de Mécanique des Sols et de Géotechnique (17ème ICSMGE), 5–9 Octobre 2009, Alexandrie, Egypte (in french), pp. 77–82 (2009b)

Dias, D., Kastner, R., Maghazi, M.: 3D simulation of slurry shield tunnelling. In: Proceedings of International Symposium on Geotechnical Aspects of Underground Construction in Soft Ground, Kusakabe, Balkema, Rotterdam, pp. 351–356 (2000)

Kasper, T., Meschke, G.: A 3D finite element simulation model for TBM tunnelling in soft ground. Int. J. Numer. Anal. Meth. Geomech. 28, 1441–1460 (2004)

Kastner, R., Ollier, C., Guibert, G.: In situ monitoring of the Lyons metro D line extension. Dans. In: Proceedings of the International Symposium on Geotechnical Aspects of Underground Construction in Soft Ground, London, UK, 15–17 avril 1996 (1996)

Lie, Z., Grasmick, J., Mooney, M.: Influence of slurry TBM parameters on ground deformation. In: ITA WTC 2015 Congress and 41st General Assembly 22–28 May, 2015, Lacroma Valamar Congress Center, Dubrovnik, Croitia (2015)

Moroto, N., Ohno, M., Fujimoto, A.: Observational control of shield tunnelling adjacent to bridge piers. Dans Underground Construction in Soft Ground, pp. 241–244. A.A. Balkema, Rotterdam (1995)

Mroueh, H., Shahrour, I.: Modélisation 3D du creusement de tunnels en site urbain. Revue Française de Génie Civil 3, 7–23 (1999). (in french)

Pantet, A.: Creusement de galeries à faible profondeur à l'aide d'un tunnelier à pression de boue. Mesures « in situ » et étude théorique du champ de déplacement. Thèse de doctorat, Institut National des Sciences Appliquées de Lyon, Lyon, France (1991)

Xie, X., Yang, Y., Ji, M.: Analysis of ground surface settlement induced by the construction of a large-diameter shield-driven tunnel in Shanghai, China. Tunnelling Undergr. Space Technol. 51(2016), 120–132 (2016)

Convergence-Confinement Method Applied to Shallow Tunnels

Yassir Naouz[✉], Latifa Ouadif, and Lahcen Bahi

Laboratory of Applied Geophysics, Geotechnics, Geology of Engineers
and Environment, Mohammadia School of Engineers,
Mohammed V University, Rabat, Morocco
naouz.yassir@gmail.com

Abstract. The assumptions of the application of the characteristic lines method, a homogeneous ground in particular, with an isotropic state of stress, made it a powerful and fast tool of pre-dimensioning the deep tunnels. For which the at-rest earth pressure coefficient is approaching 1 and subsequently the state of stress could be likened to an isotropic state. This method consists of taking into account the interaction between the soil and the structure, and thus supposes that the soil, in the hypotheses mentioned above, following a digging, discharges and converges towards the center of the excavation accompanied by an increase in the load applied on the support supposed perfectly circular and homogeneous until equilibrium.

Except these assumptions, among others when the tunnel is shallow, which is the case for almost all urban tunnels, the application of this method is no longer relevant, despite its practice! The generalization of the characteristic lines method that is proposed here consists in modeling the tunnel by a shell-type structure to which the initial ground stresses have been applied, the soil-structure interaction is considered as springs whose rigidity depends on the mechanical characteristics of the surrounding ground. The particularity of the springs is that they work as well in compression as in traction, indeed, the compression models the resistance of the ground to the thrust of the supports, as for the traction, it models a possible reduction of the stress applied by the ground on the support after its convergence.

Keywords: Convergence-Confinement method · Tunnels · Analytical methods
Shell theory · Shallow tunnels · Tunnel support · Robot structural analysis

1 Introduction

Depending on the degree of precision sought, the dimensioning of the pavements or retaining walls of tunnels can be done according to several methods. Empirical methods, like the AFTES classification (Recommendation GT7R1F2 1974), are based on feedback from projects already completed. These methods give satisfactory results for a first estimation of the necessary supports, and do not require a large amount of soil data. Semi-empirical methods, on the other hand, are based on the study of the breaking mechanisms, and thus supposes, that the breaking of a support is translated by the

© Springer Nature Switzerland AG 2019
M. Badr and A. Lotfy (Eds.): GeoMEast 2018, SUCI, pp. 53–70, 2019.
https://doi.org/10.1007/978-3-030-01884-9_5

breaking of the ground itself, consequently, the resulting internal forces are overestimated.

Soil-structure interaction methods, like the Convergence-Confinement method, involve the soil-structure behavior, consisting in the behavior of the ground strongly depending on the behavior of the pavement or retaining and vice versa. These methods, applied under certain conditions, are so precise that they give results that are very close to reality. The 3D digging effect is often modeled in the 2D model by introducing the deconfining rate λ.

Finite element numerical methods, whether in 2D or 3D, aim to get as close as possible to the actual behavior of the materials composing the model. The link between the ground and the structure is usually modeled by interfaces. At first sight, these methods make it possible to study the real behavior of the soil and give very precise results, but in the other hand they require a large amount of soil data which makes their cost high, justifying the frequent use of analytical methods.

The Convergence-Confinement method under the assumptions of an initially iso-tropic stress state, in a linearly elastic isotropic soil, has been closely treated by Panet (1995) for a perfectly circular tunnel section. Several studies that followed focused on its application on soils with perfectly elastic-plastic behavior (Corbetta et al. 1991). In the context of these very limited hypotheses, it appears that the Convergence-Confinement method is suitable for very deep tunnels when the resting force coefficient of the ground approaches 1. Excluding these assumptions, especially for shallow tunnels, or when the geometry is not necessarily circular, numerical modeling is often used which is relatively expensive.

The purpose of the present work is to propose a generalization of the Convergence-Confinement method to the case of shallow tunnels with any geometry.

2   Convergence-Confinement Method (CCM)

The meanings of the symbols used in this part are given in Table 1.

Table 1. Table of symbols

Symbol	Signification
σ_0	Initial isotropic stress in the soil
$u_r/u_e/u_{re}$	Radial/radial elastic/radial displacement at the end of the elastic phase
$E/E_b/E_s$	Soil deformation modulus/concrete/steel
ϑ/ϑ_b	Poisson coefficient of the soil/of concrete
$R/R_i/R_d$	Radius of excavation/underside of retaining/plastic radius
λ	Soil deconfinement rate
K_s	Stiffness of the support
P_a	Pressure limit of the support
σ_a	Permissible stress of the support

The Convergence-Confinement Method, also called the characteristic lines method, is based on the study of 3 curves; the ground reaction curve, the longitudinal deformation profile and the support characteristic curve:

2.1   Ground Reaction Curve (GRC)

The ground reaction curve connects the radial displacement of a point that belong to the perimeter of the excavation to an applied fictitious pressure. In the physical sense, the fictitious pressure σ^* is the minimum pressure that must be applied to the perimeter of the excavation to completely block its convergence.

The ground reaction curve depends on the soil behavior law and the failure criterion adopted in the case of non-perfectly elastic behavior.

By solving the equations of continuum mechanics in a cylindrical coordinate system (plane deformation), we can show that the ground reaction curve for an elastic linear behavior is a line (Fig. 1) passing through the two points $(0, \sigma_0)$ and $(u_e, 0)$, such that:

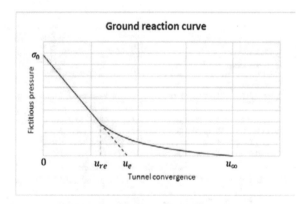

Fig. 1. Ground reaction curve

- σ_0: the initial isotropic stress
- u_e: the final radial elastic displacement of a point belonging to the perimeter of the excavation, given by the following formula:

$$u_e = \frac{1 + \vartheta}{E} . \sigma_0 . R \tag{1}$$

In the case of an elastoplastic behavior, with a Mohr-Coulomb type of plasticity criterion, two phases are distinguished; an elastic phase, during which the ground converges linearly, and a plastic phase during which the ground begins to laminate around the excavation accompanied by excessive deformation. The state of plasticization of the soil can be characterized by the plastic radius R_d, which delimits the plasticized zone, and which is linked to the displacement of the wall by the following relation:

$$\frac{u_r}{u_{re}} = \left(\frac{R_d}{R}\right)^2 \tag{2}$$

2.2 Longitudinal Deformation Profile (LDP)

The longitudinal deformation profile of the ground models the 3D aspect of the excavation. Indeed, the convergence depends, in addition to the soil characteristics and the constitutive law adopted, on the distance between the tunnel face and the section in question. The monitoring measures that have been carried out on several projects have clearly demonstrated that the soil is starting to converge upstream of the tunnel face (Tunnel of Toulon, Janin 2012). If we consider the longitudinal axis of excavation whose origin is its intersection with the tunnel face, we can distinguish three cases (Fig. 2):

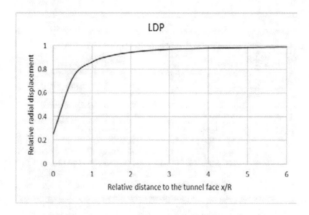

Fig. 2. Longitudinal deformation profile

- $(x \ll 0, u_{-\infty} = 0)$: This is valid for a section that is very far from the tunnel face, upstream side, where the digging effect is not yet felt.
- $(x = 0, u = u_0)$: At the face section, this means that the section has already began to converge.
- $x \gg 0, u_{+\infty} = u_\infty$: This is valid for a section that is very far from the face of the downstream side, where the convergence has totally occurred.

Although the rate of deconfinement depends on the characteristics of the soil, the expression often found in literature involves only the geometrical characteristics of the excavation and the two parameters α and m_0:

$$\lambda(x) = \alpha + (1 - \alpha)\left(1 - \left(\frac{m_0 R}{m_0 R + x}\right)^2\right) \tag{3}$$

With:

$$\sigma^* = (1 - \lambda(x)) * \sigma_0 \tag{4}$$

In the case of an elastic linear behavior, the radial displacement is connected to the final displacement by the relation:

$$u_e = \lambda(x) * u(x) \tag{5}$$

In order to determine the values of the two parameters α and m_0, we often proceed with a 3D numerical modeling that takes into account the actual parameters of the soil and the excavation. In the most advanced phases of a project, auscultation measurements can be used to make the necessary adjustments in order to calibrate the theoretical longitudinal deformation profile with the experimental one. In preliminary states, where we do not have enough soil data, we find in the literature values of 0.25 for the coefficient α and 0.75 for m_0.

2.3   Support Characteristic Curve (SCC)

When implementing the support, it opposes the convergence of the ground and applies around a uniform pressure, which depends on its radial displacement and its mechanical characteristics. Calculation of the radial displacement as a function of the applied pressure makes it possible to draw the confinement curve on a graph identical to the one of the longitudinal deformation profile.

The behavior law of a support is generally assimilated to a perfectly plastic elastic law, such that the plastic phase corresponds to its rupture.

The support characteristic curve can be defined only by the following two parameters:

- The stiffness of the support, which is defined by $K_s = P/u$
- The limiting pressure P_a

For the usual supports, these two parameters are given below:

- Chipped or projected concrete

$$K_s = \frac{E_b (R^2 - R_i^2)}{(1 + \vartheta)[(1 - 2\vartheta)R^2 + R_i^2]} \tag{6}$$

$$P_a = 0.5\sigma_a \left(1 - \frac{R_i^2}{R^2}\right) \tag{7}$$

Where σ_a is the limit stress of concrete.

- Steel arches

$$k_s = \frac{E_s S_c}{eR} \tag{8}$$

$$P_a = \frac{\sigma_a S_c}{eR} \tag{9}$$

Where:

- e is the spacing of the arches
- S_c is the section of arches

• Bolts with punctual anchoring

$$k_s = \frac{R}{e_c . e_l} \left(\frac{\pi d^2 E_s}{4l} \right) \tag{10}$$

$$P_a = \frac{T_b}{e_c e_l} \tag{11}$$

Where:

- e_c is the circumferential spacing of the bolts
- e_l is the longitudinal spacing of the bolts
- T_b is the ultimate resistant load of the bolts

The intersection between the ground reaction curve and the support characteristic curve gives the displacement and the pressure at equilibrium, and subsequently allows to dimension the support (Fig. 3).

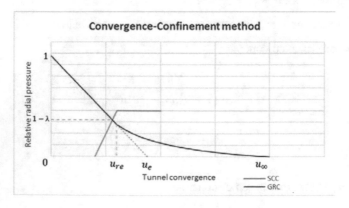

Fig. 3. Convergence-confinement method

3 Shell Theory and the Development of the Equilibrium Equations

Tunnel supports represent circular structures, with one or more arches that can ideally be assimilated to shells. Now, we will focus on the study of a support section with radius of curvature R and length $Rd\theta$. We associate to the arc the cylindrical coordinate system whose pole coincides with the center of its circle, and also the Cartesian reference (o, x, y) (Fig. 4) that we will assume to be Galilean.

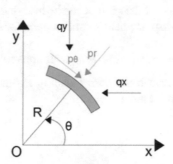

Fig. 4. External loads

The symbols used in this section are given in Table 2.

Table 2. Table of symbols

Symbol	Signification
N, M, Q	Normal effort, moment and sharpness
K_n, K_θ	Radial and tangential stiffness of the soil
u_r, u_θ	Radial and orthoradial displacement of the support
p_r, p_θ	Radial and orthoradial pressure
K	Compressibility parameter of the shell
D	Flexibility parameter of the shell
E_s, ϑ_s	Deformation modulus and Poisson coefficient of the support
A_s, I_s	Section and the Inertia of the support

3.1 Convergence-Confinement Method and the Expression of the Surcharges Applied on the Shell

In the absence of ground deconfinement, the resultant of the external loads (Fig. 4), initially applied on the section of the shell, is expressed, in cylindrical coordinates, by the following equations:

$$p_{r0} = q_x.cos^2\theta + q_y.sin^2\theta$$
$$p_{\theta 0} = \frac{q_y - qx}{2} sin(2\theta)$$

(12)

Where:

q_x and q_y are respectively the initial vertical and horizontal in-situ constraints.

When the soil declines and tends to converge towards the center of the excavation, the load p_r applied to the support decreases. Otherwise, when the support tends to push the ground, the latter opposes to this movement, and the load p_r increases.

As a result, the interaction of the ground can be introduced by the addition of a load that depends only on the displacement of the support and the characteristics of the surrounding soil.

In what follows, we opt for the following sign convention:

- Positive radial displacement if the corresponding radius increases
- Positive orthoradial displacement (Fig. 5)

Fig. 5. Sign convention for displacements

Thus, soil interaction can be formulated as follows:

$$q_r^* = \frac{u_r}{|u_r|} f(u_r)$$

(13)

The positive function f depends on the displacement of the support, the characteristics and the law of behavior of the soil.

Similarly, depending on the roughness of the interface between the ground and the support, the orthoradial displacement mobilizes a friction which can be reformulated as follows:

$$q_\theta^* = \frac{u_\theta}{|u_\theta|} g(u_\theta)$$

(14)

The positive function g depends on the orthoradial displacement of the support, the roughness of the contact surface and the characteristics of the soil. For a perfectly smooth contact, the function g is equal to zero.

The load applied by the soil has for expression:

$$\begin{cases} p_r = p_{r0} + p_r^* \\ p_\theta = p_{\theta 0} + p_\theta^* \end{cases} \tag{15}$$

For an elastic linear behavior, the radial interaction of the ground becomes equivalent to a spring of stiffness K_n such that:

$$K_n = \frac{E}{(1+\vartheta)R} \tag{16}$$

If we suppose moreover that the friction, between the support and the ground, is done without slip, the orthoradial interaction can be assimilated to a spring of stiffness K_θ.

thus:

$$\begin{cases} p_r^* = u_r K_r \\ p_\theta^* = u_\theta K_\theta \end{cases} \tag{17}$$

The load applied by the soil becomes:

$$\begin{cases} p_r = p_{r0} + K_r u_r \\ p_\theta = p_{\theta 0} + K_\theta u_\theta \end{cases} \tag{18}$$

The representation of the variation of p_r as a function of u_r makes it possible to find the ground reaction curve in the case of an elastic behavior.

For reasons of simplicity of the equations, only the case of an elastic soil will be attended.

3.2   The Third Dimension of the Excavation and the Deconfinement Rate

To take into consideration the convergence of the ground before the laying of the support, it's added to the displacement u_r in the expression of the load applied by the soil, its initial displacement u_{r0}. Thus, the expression of load becomes:

$$\begin{cases} p_r = p_{r0} + K_r(u_r + u_{r0}) \\ p_\theta = p_{\theta 0} + K_\theta u_\theta \end{cases} \tag{19}$$

The initial displacement of the soil is related to the rate of deconfinement by the relation:

$$u_{r0} = -\lambda \frac{p_{r0}}{K_n} \tag{20}$$

Thus:

$$\begin{cases} p_r = (1 - \lambda)p_{r0} + K_r u_r \\ \quad p_\theta = p_{\theta 0} + K_\theta u_\theta \end{cases} \tag{21}$$

The deconfining rate λ is determined in the same way as in $ 2.2.

3.3   Development of the Equilibrium Equations

In what follows, we adopt the following sign conventions for internal efforts:

- Positive moment if the intrados of the tunnel is compressed and the extrados is stretched
- Positive normal effort if it is a compression
- Positive sharp effort (Fig. 6)

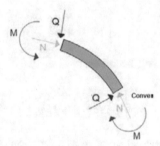

Fig. 6. Sign convention for internal efforts

By studying the static equilibrium of the section of the shell, the relationship between the internal forces and the external actions is given by the following equations:

$$\frac{dQ}{d\theta} - N + p_r.R = 0$$
$$\frac{dN}{d\theta} + Q + p_\theta.R = 0 \tag{22}$$
$$\frac{dM}{d\theta} + QR = 0$$

The behavior equations of the shell, which link the internal forces to displacements, are given below:

- For a thick shell:

$$N = -\frac{D}{R}.\left(u_r + \frac{du_\theta}{d\theta}\right) - \frac{K}{R^3}\left(u_r + \frac{d^2u_r}{d\theta^2}\right)$$
$$M = -\frac{K}{R^2}.\left(\frac{d^2u_r}{d\theta^2} + u_r\right) \tag{23}$$

– For a thin shell:

$$N = -\frac{D}{R} * \left(u_r + \frac{du_\theta}{d\theta}\right)$$
$$M = -\frac{K}{R^2} * \left(\frac{d^2 u_r}{d\theta^2}\right)$$

(24)

The compressibility parameter and the flexibility coefficient are:

• For a plane deformation state:

$$D = \frac{E_s.A_s}{1 - \vartheta_s^2}$$

(24)

$$K = \frac{E_s.I_s}{1 - \vartheta_s^2}$$

(25)

• For a plane stress state:

$$D = E_s.A_s$$

(26)

$$K = E_s.I_s$$

(27)

By injecting the expressions of the internal forces into equilibrium equations, we obtain the following equations:

– **For a thick shell:**

$$\frac{K}{DR^2}\frac{d^4 u_r}{d\theta^4} + \frac{2K}{DR^2}\frac{d^2 u_r}{d\theta^2} + \frac{du_\theta}{d\theta} + \left(\frac{K}{DR^2} + \frac{R^2}{D}K_r + 1\right)u_r + \frac{R^2}{D}(1 - \lambda)p_{r_0} = 0$$
$$-\frac{d^2 u_\theta}{d\theta^2} - \frac{du_r}{d\theta} + \frac{R^2}{D}K_\theta u_\theta + \frac{R^2}{D}p_{\theta_0} = 0$$

(28)

– **For a thin shell:**

$$\frac{K}{DR^2}\frac{d^4 u_r}{d\theta^4} + \frac{du_\theta}{d\theta} + \left(\frac{R^2}{D}K_r + 1\right)u_r + \frac{R^2}{D}(1 - \lambda)p_{r_0} = 0$$
$$\frac{K}{DR^2}\frac{d^3 u_r}{d\theta^3} - \frac{d^2 u_\theta}{d\theta^2} - \frac{du_r}{d\theta} + \frac{R^2}{D}K_\theta u_\theta + \frac{R^2}{D}p_{\theta_0} = 0$$

(29)

These equations are valid for a fixed radius R. In the case where the radius of the shell is not uniform, which is possibly the case for tunnel supports, these equations can be written on intervals of θ, on which the radius R is constant. The boundary conditions of these intervals, which reflect the continuity of the internal forces and displacements, make it possible to complete these equations in order to solve the problem (Fig. 7).

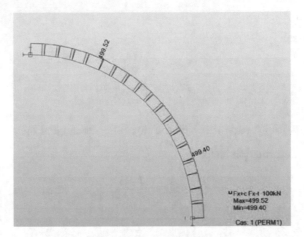

Fig. 7. Normal effort

It is noted that despite the simplifications made in terms of behavioral equations of the thin shell; negligence of the terms to the right of the thick shell equations, the order of the corresponding equations is greater than the order of the general equations for a thick shell without simplifications. Knowing that the importance of the assimilation of shells to thin shells lies in the simplification of the equations to be solved, which does not appear to be the case here, there is no interest in the study of thin shells. Therefore, the study will extend to the general case (Fig. 8).

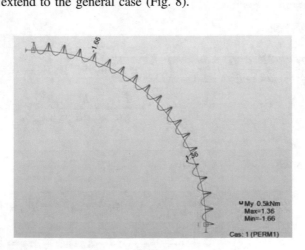

Fig. 8. Bending moment

3.4   Overview of Continuity and Boundary Conditions

The boundary conditions and the conditions of continuity between arcs make it possible to complete the differential equations, verified by the displacements of the shell, in order to find the unique solution which describes its real behavior (Fig. 9).

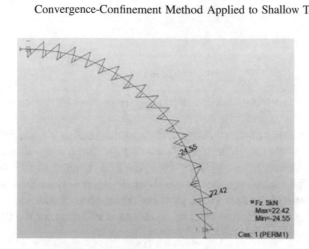

Fig. 9. Shearing force

3.4.1 Boundary Conditions

The boundary conditions depend on the support mode of both ends of the shell on the ground. These conditions can be of Dirichlet's (values), Neumann's (gradients) or Robin's (gradient/values) type. Each boundary gives 3 conditions, for a support with two boundaries, the number of conditions is 6.

3.4.2 Continuity Conditions

These conditions make it possible to describe the continuity of the internal forces and the displacements in a point of geometrical discontinuity. The geometric discontinuity may be a discontinuity of radius of the shell, or a discontinuity of angle.

3.5 Overview of the Finite Difference Method of Solving Differential Equations

The resolution of the differential equations of the shell, by the finite difference method, consists, first of all, in discretizing the derivation operators in the expressions of the equations, boundary conditions and continuity conditions thanks to the formulation of Taylor- Young, and subsequently solve, the numerical scheme thus obtained when the distance between the computation points decreases.

The discretization of the thick shell equation, by finite centered differences, is given below:

$$\frac{K}{DR^2\Delta\theta^4}U(i+2) + \left(\frac{2K}{DR^2\Delta\theta^2} - \frac{4K}{DR^2\Delta\theta^4}\right)U(i+1) + \left(-\frac{4K}{DR^2\Delta\theta^2} + \frac{6K}{DR^2\Delta\theta^4} + \frac{K}{DR^2} + \frac{R^2}{D}K_r + 1\right)U(i)$$

$$+ \left(\frac{2K}{DR^2\Delta\theta^2} - \frac{4K}{DR^2\Delta\theta^4}\right)U(i-1) + \frac{K}{DR^2\Delta\theta^4}U(i-2) + \frac{1}{2\Delta\theta}V(i+1) - \frac{1}{2\Delta\theta}V(i-1) + \frac{R^2}{D}(1-\lambda)p_{r_0}(i) = 0$$

$$- \frac{1}{2\Delta\theta}U(i+1) + \frac{1}{2\Delta\theta}U(i-1) - \frac{1}{\Delta\theta^2}V(i+1) + \left(\frac{R^2}{D}.K_\theta + \frac{2}{\Delta\theta^2}\right)V(i) - \frac{1}{\Delta\theta^2}V(i-1) + \frac{R^2}{D}.p_{\theta_0}(i) = 0$$

$$(30)$$

Where:

$$U(i) = u_r(i) \tag{31}$$

$$V(i) = u_\theta(i) \tag{32}$$

The matrix writing of the equation now discretized makes it possible to easily solve it using matrix resolution algorithms; Croute's algorithm among others.

For n computation points with $1 \le i \le n$, the first equation of the shell is valid for $3 \le i \le n - 2$, the second equation is valid for $2 \le i \le n - 1$, so we need 6 boundary conditions equations. Knowing that each boundary gives 3 additional equations, the system is of Cramer and the number of equations is sufficient for the resolution of the problem.

4 Validation of the Method Through Simple Cases and Comparison with a Numerical Calculation of Finite Element Type

In this part, we will solve the differential equation of the shell, for two types of loading; uniform radial and uniform orthoradial loading, the resolution method adopted is the finite difference method for a thick shell. The goal is to compare the solution with a numerical model of finite element modeled by Robot structural analysis.

We will study a 90° angle shell, 40 cm thick, 5 m radius, deformation modulus of 10GPa and having no geometric discontinuity. The number of computation points for the checks is equal to 50. The type of support considered for the two edges of the shell will be the sliding recessed support, i.e. locked rotation, blocked orthoradial displacement and unblocked radial displacement.

As for the characteristics of the soil, we consider a linear elastic law with a deformation modulus of 100 MPa and a Poisson coefficient of 0,25. The orthoradial stiffness is taken equal to half of the radial stiffness.

For the model finite elements on Robot, the shell will be assimilated to a succession of beams recessed relative to each other, the interaction on the ground is assimilated to springs of the same radial and orthoradial stiffness as in our calculation. Every other calculation assumptions related to the geometric and physical characteristics of the shell and the modes of support are conserved.

The calculation assumptions are summarized in the Table 3 below:

4.1 Radial Pressure of 100 KPa

In this part, we will consider the shell subjected to a uniform radial pressure of 100 kPa.

To evaluate the impact of soil structure interaction on the results. Two calculations were made. The first calculation is made for a uniform pressure of 100 kPa, independent of the deformation of the support, which amounts to neglecting the soil

Table 3. Table of parameters

Parameter	Value
Thickness of the shell	0.4 m
Radius of the shell	5 m
Angle of the shell	90°
Modulus deformation of the shell	10 GPa
Ground deformation modulus	100 MPa
Poisson's coefficient of soil	0.25
Radial pressure	100 kPa
Orthoradial pressure	100 kPa
Orthoradial stiffness K_θ	0.5* K_r
Number of calculation points	50
Support mode	Sliding recessed
Deconfinement rate	0

deformation modulus. The second calculation is done with a soil deformation modulus of 100 MPa.

The results of the first calculation are given in the following table:

θ	0	$\pi/4$	$\pi/2$
$u_\theta(m)$	0	0	0
$u_r(m)$	−0.006	−0.006	−0.006
$M(KN.m)$	1.3	1.3	1.3
$N(KN)$	500	500	500
$Q(KN)$	0	0	0

Robot results:

It is found that the normal effort found by Robot equals the calculated normal effort. As for the moment and the shearing force, we note that Robot's results fluctuate between two extreme values, and that the peaks of these internal forces are situated at the points of the connections between the different beam elements; this is due to the assimilation of the support to a series of beams. In the model of the shell, we see that the internal forces are constant.

The second calculation with a soil deformation modulus of 100 MPa gives the following results:

	$N(KN)$	$M(KN.m)$	$Q(KN)$
Calculation	440KN	1.1	0
Robot	440KN	1.1; −1.1	11; −12

The calculated normal effort equals the results achieved by Robot, we also note that the calculated moment value corresponds to the maximum value of the moment found by Robot. However, the value of the shear force given by Robot differs a lot from the null value found by the calculation.

4.2   Orthoradial Pressure of 100 KPa

In this part, we will consider the shell subjected to a uniform orthoradial pressure of 100 kPa. Following the same pattern, two calculations have been made; a calculation neglecting the soil deformation modulus, and a calculation with a 100 MPa modulus.

The results of the two calculations are given in the tables below.

- Calculation with a soil deformation modulus of zero

The internal forces varying linearly between the two extreme values at the supports, given in the following table:

	$N(KN)$	$M(KN.m)$
Calculation	450; −450	350; −350
Robot	410; −470	476; −493

- Calculation with a modulus of 100 MPa

	$N(KN)$	$M(KN.m)$
Calculation	350; −350	20; −20
Robot	310; −390	30; −30

It is found that the normal efforts found by Robot are consistent with those calculated. On the other hand, the values of the moments differ in the order of 70%. It is also noted that taking into consideration the soil-structure interaction makes it possible to have lower internal forces.

5   Discussion

The comparisons between the calculation results and those of Robot, allowed us to validate the calculation approach. Indeed, although a thick shell modeling differs a lot from continuous beam modeling, like the one used in Robot structural analysis, the results towards normal efforts are almost identical.

The differences of moments values are due to the parameters of the numerical computation, in addition to the way in which the support for the two approaches are modeled. The number of calculation points for the finite difference method and the number of finite elements used in Robot greatly influenced the results.

The study of circular support by the theory of shells gives more precise results than those of Robot. Indeed, the sharp angle variation between two beam-type elements creates a stress concentration, this can give rise to significant internal forces.

6 Conclusion

The formulation of the equations of the behavior of the thick shells, by integrating the interaction of the ground and the 3D effect of the excavation by means of the rate of deconfinement, allowed us to widen the field of application of the Convergence-Confinement method to the case of an anisotropic stress state with a section of geometry that is not necessarily circular. The equation of the shell has been established in the hypotheses of an elastic behavior. For an elastoplastic constitutive law, it is necessary to adapt the functions f and g in the calculation of the load applied by the soil and considering the criterion of plasticity and the deformability of the soil.

The comparisons with Robot computation, made it possible to validate the calculation approach on simple loading examples; uniform radial and orthoradial. The normal efforts found by the calculation come very close to the efforts found by Robot. As for the shearing force and the bending moment, there is a significant difference in the results, a finite element model modeling the shell by a succession of beam elements gives overestimated efforts at the level of the connections between the different elements, on the other hand the calculation by the shell theory gives more realistic internal forces.

This article aims to improve the confinement convergence method by adapting it to the context of shallow tunnels with a state of anisotropic stress. This methodology is used to dimension tunnels in preliminary studies and has the advantage of being more precise than the empirical and semi-empirical methods, it's also faster and cheaper than the sophisticated numerical methods that are rather used in advanced project studies.

References

AFTES: Recommandations pour l'emploi de la méthode Convergence-Confinement. Tunn Ouvrages Souterr 170 (2002)

Autodesk Robot Structural Analysis. Imperial Getting Started Guide (2010)

Carranza-Torres, C., Labuz, J.: Class notes on underground excavations in rock. Topic 6. USA: Department of civil engineering, University of Minessota (2006)

Carranza-Torres, C., Diedrichs, M.: Tunneling and underground space technology **24**, 506–532 (2009)

Champagne de Labriolle, G .: Amélioration des méthodes analytique basées sur des concepts simples pour le dimensionnement des tunnels superficiels et profonds en sol meuble (2016)

Corbetta, F., Bernaud, D., Nguyen Minh, D.: Contribution à la méthode Convergence-Confinement par le principe de la similitude. Rev. Fr. Geotech. **54** (1991)

Faihurst, C., Carranza-Torres, C.: Closing the circle. In: Proceedings of the 50th Annual Geotechnical Engineering Conference (2002)

Flûge, W.: Stresses in Shells. Springer, New York (1960)

Janin, J.P.: Tunnels en milieu urbain, prévisions des tassements avec prise en compte des effets des pré-soutènements (2002)

Panet, M.: Le calcul des tunnels par la méthode de convergence Convergence-Confinement. Presses de l'ENPC, Paris (1995)

Mousivand, M., Maleki, M., Nekooei, M., Mansouri, M.: Application of convergence-confinement method in analysis of shallow non-circular tunnels (2017)

Panet, M.: La mécanique des roches appliquée aux ouvrages de génie civil. Presses de l'Ecole Nationale des Ponts et Chaussées, 235 p (1976)

Panet, M.: La stabilité des ouvrages souterrains: soutènement et revêtement. Laboratoire des Ponts et Chaussées, Rapport de recherche n 28, 32 p (1973)

Panet, M., Guenot, A.: Analysis of convergence behind the face of a tunnel. In: International Symposium on Tunneling, vol. 82, Brighton (1982)

Poromechanical Behavior Analysis of an Underground Cavity Below Runways Under the Dynamic Cyclic Action of Landing Gear on Complex Geotechnical Conditions

Mohamed Chikhaoui[✉] and Ammar Nechnech

LEEGO Laboratory, Faculty of the Civil Engineering, University of Sciences and
Technology Houari Boumediene (USTHB), Bab Ezzouar, Algeria
mch_gcgl6@yahoo.ca

Abstract. This paper deals the analysis of the behavior of an underground
saline cavity under the cyclic load due to traffic landing gear from the Boeing
747-400 airplane using the pseudo static approach. This cavity was located at a
depth of 2.5 m from the free surface of the second runway at Es-Senia airport in
Oran (Algeria). Where, this airstrip is built in very complex geotechnical conditions, which several cavities are formed by dissolution of gypsum, have been
observed during the geological investigations. In-situ geotechnical studies and
experimental tests carried out in the laboratory have permitted to identify the
hydromechanical behavior parameters of the great Sebkha soil of Oran city. The
Oran airport base is composed of a mixture of clay and silty gypsum sand with
an overconsolidated behavior.

This article focuses on numerical modeling of the second airstrip of Oran
airport. This modeling was performed using the finite difference software Flac^{3D}
(v.3.1), while assuming that a spherical subterranean cavity is located 2.5 m
deep. The obtained results have allowed representing the real behavior of the
underground cavity and that of the runway under the action rear main landing
gear of the Boeing 747-400 considering only the poromechanical behavior in
this part of study.

1 Introduction

Recently, many studies have been devoted to analyze the behavior of flexible pavements via numerical finite elements and finite differences models able to accurately
determine with precision the damage of the subsoil and that of the aeronautical
pavements caused by the impact of the rear landing gear from the heavy aircrafts with
several wheels from the heavy aircrafts.

Although, the mechanics researchers are interested in the internal friction of the
tires as well as the improvement of the mechanisms of landing gear. While, geomechanics are interested in the dynamic signal generated by the impact of dynamic loads
on the ground, more precisely the behavior of these soils under external constraints,
such as the impact of landing gear on the track and especially the ground support, or
sometimes internal actions.

© Springer Nature Switzerland AG 2019
M. Badr and A. Lotfy (Eds.): GeoMEast 2018, SUCI, pp. 71–80, 2019.
https://doi.org/10.1007/978-3-030-01884-9_6

For example, Alvappillai et al. (1992) proposed an algorithm based on the finite element method. This program is used to analyze multiple dynamic responses under the action of rolling mobile aircraft loads on concrete pavements with joints. Pan and Atluri (1995) discussed the behavior of runways with respect to the dynamic effect caused by Boeing 747-200B tires in the take-off phase only.

In addition, the study of Pan and Atluri (1995) shows that running the aircraft at a higher speed induces a large deflection, where the deflections produced degrade the roadway in the long-term. As a result, they can cause an increased stresses in the coating and affect the service life due to fatigue of the pavement structure. Moreover, Besselink (2000) has produced a numerical model for simulating "shimmy" lateral oscillations in the main landing gear of a Boeing 747-400. The data measured by both Besselink (2000) and after Khapane (2004) shows that the acceleration of aircraft tires from rotation to zero speed up to a free speed of about 0,1 s of touching tires on the ground (impact). Nevertheless, Persson (2006) was interested in the friction forces generated between the surface of the tire contact and the wearing course. On the other hand, Broutin (2010) mentioned in his thesis, that only the materials affected by the loading time are the surface layers and exhibit viscoelastic behaviors (or as a function of time). As for Leonardi (2014) it is interfered with the case of repeated cycles of the impact of an Airbus main landing gear wheel (A321) on a flexible pavement in order to predict the distribution of the permanent deformation of the wearing course.

Alroqi and Wang (2015) studied the impact of the Boeing 747-400 aircraft landing by using a single-wheeled RSE physical model of its main landing gear for the purpose make tire life corrections by reducing abrasive skidding between aircraft tires and contact surfaces of runways immediately after touchdown.

While the cyclic impact itself aircraft produces forces that can reach the body of pavement or even the ground support. These oscillating stresses lead to a dissipation of energy and will cause the surface layer and the pavement body to degrade by repeated movement (heating of the tire, rutting and subsidence of the support ground).

In the north of the great Sebkha of Oran, the runway Es-Senia airport is built in very complex geotechnical conditions, which several cavities are formed by dissolution of gypsum (Chikhaoui et al. 2011, 2013, 2015 and 2017). The presence of this kind of problem can cause considerable damage. For that, the problem treated in this works concerns the analysis of the behavior of an underground saline cavity located at a depth of 2.5 m from the free surface of the second runway at Es-Senia airport in Oran (Algeria) under the cyclic load due to traffic landing gear from the Boeing 747-400 airplane using the pseudo static approach.

In fact, this article aims to perform numerical modeling of the Oran airport runway, taking into account the presence of a spherical underground cavity located at 2.5 m depth, using Flac^{3D} (v. 3.1) software. This numerical modeling let to study the poromechanical behavior of a saline underground cavity and the stability of the runway under the cyclic effect from the Boeing 747-400 landing gear via pseudo dynamic approach.

This analysis was carried out considering the poromechanical properties subgrade as well as those of the airstrip. These parameters were characterized by in-situ tests, namely SPT and pressiometer tests, as well as by laboratory tests conducted on reconstituted soil samples. Besides, according to the consolidated undrained

(CU) triaxial tests results, described by Chikhaoui et al. (2017). The ground of the runway is principally composed of a sandy and gypseous clay loam with a thickness varying from 02 to 24 m. Moreover, the soil behavior is overconsolidated kind.

For studying the stability or the collapse of runways at long-term, the cavities size variations was taken into consideration and the dynamic interaction between the wheels of landing gear and specific geotechnical conditions was considered also.

2 Materials and Methods

2.1 Pavement Model

In order to take into consideration the effect of landing gear via $Flac^{3D}$ in the presence of underground cavity by considering only the poromechanical behavior. A symmetrical modeling was established considering only the half of the runway, taking also into account the half of the Boeing 747-400 rear main undercarriage (Fig. 1).

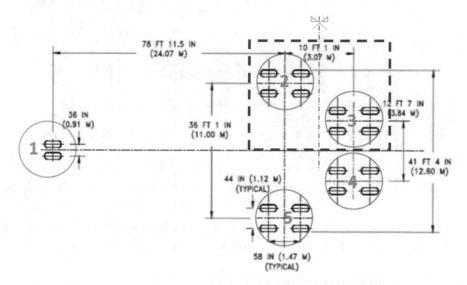

Fig. 1. Half of the Boeing 747-400's rear main landing gear taken in the modeling

For that, the adequate geometry representative of the impact phase and the first phases of moving of landing gear in the presence of an underground cavity supposed existing in the subsoil of the track has the following dimensions: 15 m in the X direction, 22.5 m in the Y direction and 30 m of thickness taken in the Z direction. The underground cavity is a spherical size with diameter $\varphi = 1.5$ m located at a depth of 2.5 m in the first layer of the basement of the track. The final model chosen is composed by four layers of grounds including the under-layers composing the body of roadway (Fig. 2).

The parameters of the upper layers of the carriageway are shown in Table 1.

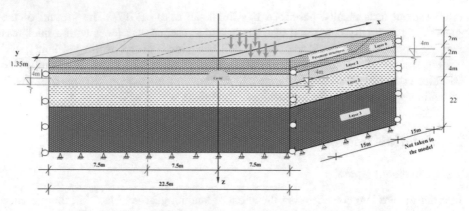

Fig. 2. The subsoil numerical modeling of the Es-Senia Airport

Table 1. The substructure runway parameters input on Flac^{3D} code.

		Naming	Linking type	Thickness	γ Kg/m^3	E (Pa)	ν
The surface layer	Top layer	Bituminous concrete (BB)	Attached	8 cm	2350	4,32 10^{+09}	0,35
	Bonding layer	Base asphalt (GB-2)	Attached	12 cm	2350	7,50 10^{+09}	0,35
The sitting layer	Base layer	"As dug" gravel (GNT-B2)	Attached	35 cm	2250	4,80 10^{+08}	0,35
	Foundation layer	Tuff 0/40	Attached	60 cm	2200	3,80 10^{+08}	0,35
The subsoil	Subgrade layer	Tuff 0/80	Attached	20 cm	2200	3,80 10^{+08}	0,35

The hydromechanical parameters used for this modeling were identified through the Chikhaoui M. (2018) thesis and the Chikhaoui et al. (2017) article (Table 2).

2.2 Boundaries Conditions and Hypothesis

The pseudo dynamic boundary conditions of displacement are necessary to represent the far-field behavior. Appropriate displacement boundary conditions allow representing body displacements in the limits of the model. These boundaries have been defined as the vertical geometric lines for which the abscissa (x) is equal to the smallest or the largest, for which the abscissa X and the ordinate Y of the profile (x = 0 m, x = 15 m y = -7.5 m, x = 22.5 m) are blocked in the horizontal direction (Ux = Uy 0). Moreover, the geometric lines for which the direction (Z) at depth is equal to the smallest value of the model are completely fully blocked (Uz = 0), except at the surface which is considered free in order to follow the real deformed form of the profile (Fig. 2).

Table 2. The subsoil hydromechanical parameters input on Flac^{3D} code.

	Layer 0	Layer 1	Layer 2	Layer 3
Depth (m)	0 to 2 m	2 to 4 m	4 to 8 m	8 to 24 m
Permeability coefficient K_{10} (m/s)	$1.25\ 10^{-06}$	$1.46\ 10^{-07}$	$1.23\ 10^{-07}$	$8.34\ 10^{-08}$
n (%)	32.62%	34.99%	30.96%	30.15%
γ_h (Kg/m^3)	1899	1910	1843	1899
γ_d (Kg/m^3)	1789	1598	1555	1499
E (Pa)	$9.60\ 10^{-7}$	$9.10\ 10^{-7}$	$1.50\ 10^{-8}$	$2.01\ 10^{-8}$
v	0,37	0,4	0,36	0,26
C' (KPa)	5.13	58.95	105.18	28.13
φ' (°)	38.16	38.45	34.81	36.16
Ψ (°)	6.87	7.14	6.86	8.38

Generally, the technique of approaching an aircraft is based on maintaining a fixed speed to reach 50 feet (15 m) on the threshold of the runway. The horizontal velocity in relation to the landing speed is equal to the approach speed of Boeing (2011), 80.78 m/s at the impact minus 5.14 m/ s from the beginning of the deceleration (Ochi and Kanai 1999), resulting in a horizontal landing speed of 75.6 m/s after impact. Where, a deceleration of the order of 4 to 6 ft/s (1.219 to 1.829 m/s) is operated gradually until the touch of the tires on the ground. So, the fuselage principle is used to reduce the vertical speed and to make a soft landing to reduce the landing stroke distance.

According to the manufacturer's data (Boeing 2011), the beginning rotation time of the tires after impact is 0.1 s for a Boeing 747-400 according to Khapane (2004) and Boeing (2002) with a progressive deceleration, so the activation of the first braking phase is usually after 1 s. This was reported in the article of Khapane (2004) as well as the article of Leonardi (2014) for the case of Airbus (A321).

While, the time sensitive to the intensity of the dynamic impact is considered important in the interval of 7 to 35 ms according to Broutin (2010), STAC (2014) Broutin and Sadoun (2016).

In this work, it was assumed that:

- The critical times involved in the calculations were set as follows: t = 0; 0.007; 0.009; 0.01; 0.012; 0.015; 0.018; 0.019; 0.02; 0025; 0.035; 0.05; 0.08; 0.09 and 0.095 s;
- For that, only times from 0 to 0.05 s were considered in this article.
- Where, the landing speed at impact was set at 80.78 m/s;
- The sliding friction phases of the tires on the roadway were not taken in the model.

2.3   Validation

The results of stresses and displacements responses obtained, are almost similar to the Boussinesq (1886) representation, relating to the variation of the stresses at the surface and at the depth of the ground. Where, it has been found that the stresses due to the impact of the rear main undercarriage are more important at the surface and begin to be attenuated in depth. In addition, the effect of the impact of the undercarriage has no effect from more than 5 m under the basement of the track without cavity (intact model) and from 8 m deep for the case of track in the presence of cavity (Fig. 3).

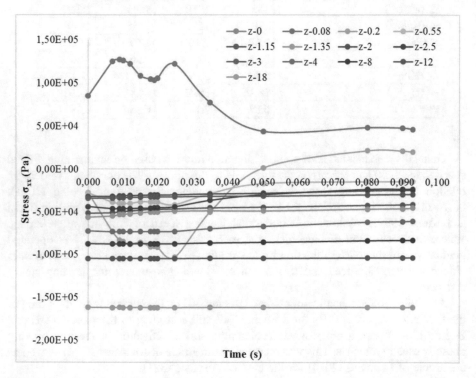

Fig. 3. Pseudo-static stress variation σxx example in the plan (x, y) according to different increments considered times following the impact in the position (x, y) = (1.5, 0) for different depths z.

Furthermore, it has been found that there are two critical times recorded in the two directions X (lateral direction) and Y (direction of movement of the aircraft). Due to the impact of the main undercarriage on the surface of the runway, the obtained critical times are t = 0.035 s and t = 0.025 s (Fig. 4). These critical times are obtained respectively in the lateral direction (X-axis) and in the direction of the aircraft's movements (Y-axis), with a X displacement (X_{deplts}) of the order of ($-3.39 \ 10^{-4}$ m at t = 0.035 s and $-4.45 \ 10^{-5}$ m at t = 0.025 s recorded at x = 5.15 m corresponding to the second landing bogie. But for the Y displacement (Y_{deplts}) it is about ($-8.94 \ 10^{-5}$ m at t = 0.035 s and $1.12 \ 10^{-4}$ m at t = 0.025 s) recorded at x = 1.5 m corresponding to the first landing bogie.

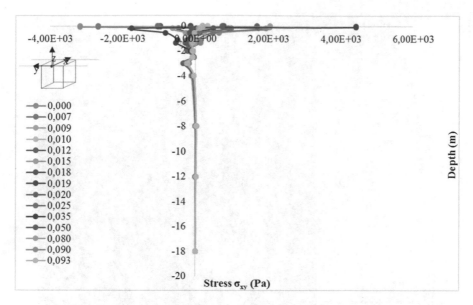

Fig. 4. Pseudo-static shear stress variation σxy example in the plan (y, z) according to different depths z for different increments considered times following the impact in the position (x, y) = (0, 0).

These displacements relate to the constraints recorded in the two directions (X, Y) and which are; σ_{xx} stresses is ($-8.99 \ 10^{+5}$ m at t = 0.035 s and $-9.28 \ 10^{+5}$ m at t = 0.025 s), then for σ_{yy} stresses is about ($-1.37 \ 10^{+6}$ m at t = 0.035 s and $-1.39 \ 10^{+6}$ m at t = 0.025 s). This results joins the findings of (Broutin 2010, 2014 and STAC 2014).

3   Results and Discussions

The variations of the stress tensors are much more important also to the right of the contact area of the tires on the roadway. Those stress are more accumulated rearward the tires contact with the roadway (cavity side), as the landing gear progresses. However, tensors of recorded constraints are amply larger at t = 0.035 s and t = 0.025 s collected respectively in the X and Y directions.

A slight tension is seen at the sidewalls of the cavity in the moving direction of the aircraft (from right to left). This tension (plasticization) increases as the aircraft progresses until its attenuation at t = 0.08 s, it is maximum at t = 0.035 s. In addition, more the diameter of the cavity is important, further, the plasticizing is also important. For example, the figure below (Fig. 5) shows the state of the plasticization at t = 0.035 for the cavity diameters φ = 1.5 m. This plasticization is consumed in the first cycle moving of the aircraft or the landing gear. That is, as soon as the 2nd landing cycle is activated, the cavity remains in its previous plasticizing state without any change in the shape of the cavity. Although, a slight displacement is recorded of about 1/10000 cm for each passage (for the 2nd cycle and more, or even for 40 landing cycles).

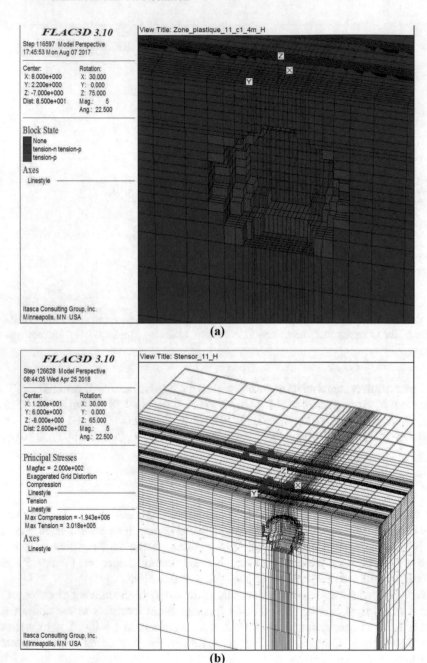

Fig. 5. (a). Plasticization of the cavity 1.5 m size at t = 0.035 s in the left and the correspondent position of landing gear on the runway (b).

4 Conclusions

In this work, the results of the three-dimensional numerical modeling via $Flac^{3D}$ software, relating to the effect of the rear main landing gear of the Boeing 747-400 on the airstrip of Es-Senia at Oran in the presence of a supposed spherical cavity existing at -2.5 m depth, were presented taking into account only the poromechanical behavior.

The pseudo-static approach adopted to establish this study has been highlighted by the results obtained and is similar to those found by different researchers working on this research axis. The advancement of the landing gear on the roadway, led to conclude, that it can causes accumulations due to the relatively high concentrations of the stresses and displacements to generate finally the fatigue weakening of the cavity walls and probably their collapse under the repeated landing action.

The obtained results are really encouraging for conducting other simulations to combine other behavior, considering also the soil behavior in the case of reinforcement of existing cavities by using several techniques of reinforcement under the landing runways.

Acknowledgments. The authors would like to thank the head civil engineering team of the PRISME laboratory at the University of Orleans in France for allowing us to use their Version software $Flac^{3D}$ (V3.1).

References

Alroqi, A., Wang, W.: Comparison of Aircraft Tire Wear with Initial Wheel Rotational Speed. Int. J. Aviat. Aeronaut. Aerosp. (2015). https://doi.org/10.15394/ijaaa.2015.1043

Alvappillai, A., Zaman, M., Laguros, J.: Finite element algorithm for jointed concrete pavements subjected to moving aircraft. Comput. Geotech. **14**(3), 121–147 (1992). https://doi.org/10.1016/0266-352X(92)90030-W

Besselink, I.J.: Shimmy of aircraft main landing gears. Delft: Technische Universiteit Delft. Ph. D. thesis. Holond (2000). Available from http://www.tue.nl/en/publication/ep/p/d/ep-uid/227775/

Boeing, Boeing 747-400. Airplane Characteristics for Airport Planning. Boeing.USA (2002)

Boeing, Commercial Airplane Co. Approach speeds for Boeing airplanes. Boeing Airplane (2011). Available from http://www.boeing.com/assets/pdf/commercial/airports/faqs/arcand approachspeeds.pdf

Broutin, M.: Assessment of flexible airfield pavements using Heavy Weight Deflectometers. Development of a FEM dynamical time-domain analysis for the backcalculation of structural properties. Structures and Materials, ENPC-LCPC. France (2010)

Broutin, M., Sadoun, A.: Advanced Modelling for Rigid Pavement Assessment Using HWD. Transp. Res. Proc. **14**(Supplement C), 3572–3581 (2016). https://doi.org/10.1016/j.trpro.2016.05.425

Chikhaoui, M.: La prise en compte du couplage Hydrothermomécanique dans l'effondrement des pistes d'atterrissage construites sur un sol salin. Géotechnique, Université des Sciences et de Technologies - Houari Boumediènne, Alger. Algérie (2018)

Chikhaoui, M., Belayachi, N., Nechnech, A., Hoxha, D.: Experimental Characterization of the Hydromechanical Properties of the Gypsum Soil of Sebkha of Oran. Period. Polytech. Civil Eng. (2017). https://doi.org/10.3311/PPci.10468

Chikhaoui, M., Nechnech, A., Haddadi, S., Ait-Mokhtar, K.: Contribution a la prédiction des effondrements des pistes d'aérodromes construites sur des sols salins. In: Gestion des risques d'exploitation routière, du XXIVe Congrès mondial de la Route de Mexico 2011 du 26 au 30 Sept 2011, pp. 1–10 Mexico (2011)

Chikhaoui, M., Nechnech, A., Hoxha, D., Moussa, K.: Contribution to the Behavior Study and Collapse Risk of Underground Cavities in Highly Saline Geological Formations. In: Lollino, G. et al. Engineering Geology for Society and Territory – Vol. 6, pp. 393–397. Springer International Publishing, Cham. Available from http://link.springer.com/10.1007/978-3-319-09060-3_68 Accessed 2 July 2016 *2015)

Chikhaoui, M., Nechnech, A., Moussa, K.: Contribution to the study of coupling hydro – thermo mechanical for the prediction of collapses of airfield runways built on saline soil. In: 14th REAAA Conference 2013. Road Engineering Association of Malaysia (REAM), pp. 638–648. From 26–28 March 2013. REAAA, Kuala Lumpur – Malaysia (2013)

Khapane, P. 2004. Simulation of Aircraft Landing Gear Dynamics Using Flexible Multibody Dynamics Methods in Simpack. In ICAS 2004,. Yokohama, Japan, 29 August - 3 September 2004. Available from www.icas.org/ICAS_ARCHIVE/ICAS2006/PAPERS/600.PDF

Leonardi, G.: Finite element analysis of airfield flexible pavement. Arch. Civil Eng. 60(3) (2014). https://doi.org/10.2478/ace-2014-0022

Ochi, Y., Kanai, K.: . Automatic approach and landing for propulsion controlled aircraft by H/sub ∞/ control, pp. 997–1002. IEEE (1999). https://doi.org/10.1109/cca.1999.800951

Pan, G., Atluri, S.N.: Dynamic response of finite sized elastic runways subjected to moving loads: A coupled BEM/FEM approach. Int. J. Numer. Methods Eng. 38(18), 3143–3166 (1995). https://doi.org/10.1002/nme.1620381808

Persson, B.N.J.: Rubber friction: role of the flash temperature. J. Phys.: Condens. Matter 18(32), 7789–7823 (2006). https://doi.org/10.1088/0953-8984/18/32/025

STAC, Ascultation des chaussées souples aéronautiques au HWD (Heavy Weight Deflectometer). Direction général de l'Aviation civile. Service technique de l'Aviation civile, 12 Feb 2014 (2014)

Earth Pressure Distribution on Rigid Pipes Overlain by TDA Inclusion

Mohamed A. Meguid[⊠]

Civil Engineering and Applied Mechanics, McGill University, Québec, Canada
mohamed.meguid@mcgill.ca

Abstract. Rigid pipes installed under high embankments are often installed using the induced trench technique by introducing a compressible zone above the pipe. A new application of tire-derived aggregate (TDA) for induced trench rigid pipes is evaluated in this study. An experimental investigation that has been conducted to measure the earth pressure distribution on a rigid pipe buried in granular material and backfilled with TDA is presented in this study. A setup has been designed and built to allow for the installation of an instrumented pipe in granular material and measuring the contact pressure acting on the pipe wall. Results show that the use of TDA as a compressible material can provide similar beneficial effects to rigid pipes compared to expanded polystyrene (EPS) geofoam and other commonly used soft inclusions. The earth pressure is found to decrease significantly all around the pipe compared to the conventional installation with aggregate backfill. The average measured earth pressure above the crown of the pipe was found to be as low as 30% of the overburden pressure for installations with granular backfill material. The pressure at the invert also decreased by about 75% with the introduction of the soft TDA zone above the pipe.

Keywords: Soil-structure interaction · Tire-derived aggregate
Buried pipes · Induced trench installation · Soil arching · Contact pressure measurement

1 Introduction

The construction of buried pipes and culverts under high embankment fills can impose significant earth loads on these structures depending on the installation condition. The vertical earth pressure on a rigid conduit installed under earth embankment is generally greater than the weight of the soil above the structure because of negative arching. To reduce the vertical earth pressure on rigid pipes and culverts, the induced trench installation (ITI) method introduced by Marston (1922, 1930) can be used. In this method, the loads are redistributed around the buried structure by introducing a compressible material above the top of the pipe to promote positive arching as depicted in Fig. 1.

© Springer Nature Switzerland AG 2019
M. Badr and A. Lotfy (Eds.): GeoMEast 2018, SUCI, pp. 81–93, 2019.
https://doi.org/10.1007/978-3-030-01884-9_7

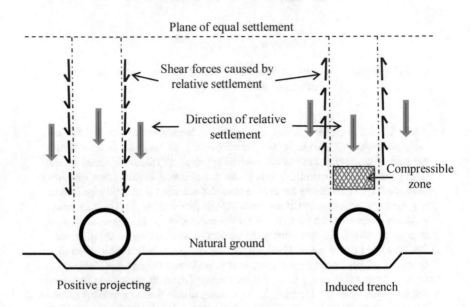

Fig. 1. Schematic of the induced trench method

The ITI installation method for rigid conduits buried under high embankments dates back to the early 1900s. Researchers studied the relevant soil-structure interaction using experimental testing and field instrumentation (e.g. Sladen and Oswell 1988; Vaslestad et al. 1993; Liedberg 1997; Sun et al. 2011; Oshati et al. 2012; Meguid et al. (2017a, b); Meguid and Youssef 2018) as well as numerical modelling (Kim and Yoo 2002; Kang et al. 2008; McGuigan and Valsangkar 2010; Meguid and Hussein 2017) to help understand the method and to address uncertainties associated with this design approach. The material used for the compressible zone mostly consisted of sawdust, wood chips, or EPS geofoam.

Tire-derived aggregate (TDA) is an engineered product made by cutting waste tires into small pieces (25–305 mm) that can be used to replace natural aggregates in civil engineering applications (Humphery 2004). Extensive research has been done over the past two decades (e.g. Bosscher et al. 1992; Drescher and Newcomb 1994; Tweedie et al. 1998; Strenk et al. 2007; Mills and McGinn 2010) to examine the performance of TDA under different loading conditions. These studies demonstrated that TDA is a suitable fill material with engineering properties that are comparable to conventional aggregates. Applications include the construction of large embankments, bridge abutment backfill, retaining walls and pavement subgrade.

TDA is generally classified, based on size and gradation, into two types: Type-A, with a maximum particle size of 75 mm, and Type-B with a maximum particle size of 305 mm (ASTM D 6270). Direct shear tests performed on Type-A TDA samples

(Humphery 2008; Xiao et al. 2015) suggested a friction angle that ranges from 36° to about 39° for TDA-sand interface.

In this study, an investigation into the contact pressure distribution on buried pipes is conducted using laboratory experiments. The experimental work involves a thick-walled PVC pipe that is instrumented with tactile pressure sensors and buried in granular material while a distributed load is applied at the surface. The tactile sensors used in this study are adapted from the robotics industry and have been successfully used in geotechnical engineering applications to measure the distribution of normal stresses in granular soils (e.g. Paikowsky et al. 2000; Springman et al. 2002; Ahmed et al. 2015). Due to their limited thickness, tactile sensors possess favourable charac-teristics with respect to aspect ratio and stiffness over the conventional load cells. In addition, being flexible enables shaping the sensing pads to cover curved surfaces, hence suitable for cylindrical shape structures. Meguid et al. (2017a, b) and Meguid and Youssef (2018) used tactile (TactArray) sensors to measure contact pressure dis-tributions on both circular as well as square shaped structures.

The measured pressures are initially presented for a benchmark condition where only granular backfill material is used and the results are compared with Hoeg's analytical solutions (Hoeg 1968). A series of experiments are then performed by incrementally increasing the surface pressure for two backfill conditions: (1) the pipe is backfilled with only granular material; and (2) a layer of TDA is introduced above the pipe. The experimental setup and test procedure are presented in the next section, followed by the pressure results of the two-backfill conditions.

2   Experimental Setup

The experimental setup consists of a thick-walled pipe embedded in granular backfill material contained in a test chamber. The pipe is instrumented using tactile sensing pads - wrapped around its outer perimeter - covering the area near the middle third of the pipe length. A universal MTS testing machine with a capacity of 2650 kN is used to apply distributed load, utilizing a rigid steel platform. A detailed description of the experimental setup is given below.

2.1   Test Chamber

The test chamber used in the experiments is shown schematically in Fig. 2. The dimensions (1.4 m × 1.0 m × 0.45 m) are selected in order to represent a two-dimensional loading condition. The rigid walls are placed far from the pipe to minimize boundary effects; wherein the distance from the outer perimeter of the pipe to the sidewalls of the tank is more than 4 times the pipe diameter (Leonard and Roy 1976). All steel wall surfaces are epoxy coated and covered with two plastic sheets separated using a thin layer of grease to minimize friction between the backfill material and the rigid walls.

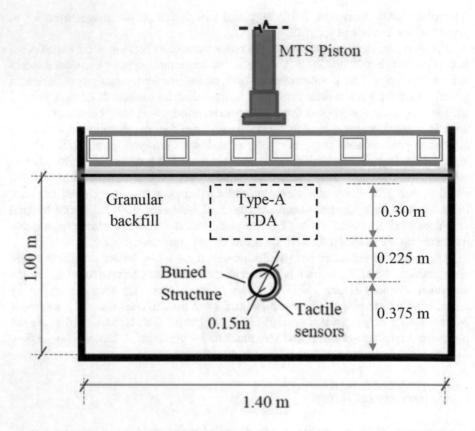

Fig. 2. Schematic of the test chamber

The pipe used for this study (15 cm in outer diameter and 1 cm in wall thickness) is instrumented using two custom-made sensing pads installed directly on the pipe wall. Each sensing pad contains 255 square-shaped sensors with pressure ranging from 0 to 140 kPa. The sensors are protected from backfill abrasion by covering the instrumented pipe with a thin layer of stiff rubber sheet.

Two LVDTs are installed orthogonally inside the pipe to confirm the rigid pipe assumption. The maximum diameter change, at surface pressure of 100 kPa, was found to be less than 0.04 mm, which is considered insignificant.

2.2   Material Properties and Testing Procedure

The dry sandy gravel backfill used in this study has an average unit weight of 16.3 kN/m^3. Sieve analysis, conducted on selected samples, indicated a coarse-grained material with 77% gravel and 23% sand. The friction angle of the backfill - determined using direct shear tests - is found to be 47°. The soil was placed and tamped in layers to form a dense base bedding material below the pipe. This was achieved by raining the sand from a constant height into the chamber in layers. From the base up to the pipe invert, the soil was placed in three layers 100 mm in height. Each layer was first graded to level the surface then tamped using a steel plate with a long handle to reach the base of the test chamber. The sand placement continued up to the pipe invert. Above the invert, the rained sand was gently pushed around the pipe up to the crown to ensure full contact between the sand and the pipe. Then, another layer of sand was added to cover the pipe. The remaining sand required to reach the height above the crown was placed and graded to minimize damage to the pressure sensors. The instrumented pipe was placed over a thin film of sand to improve the contact between the soil and the pipe. Density of the backfill was measured in the rigid box using density cups placed at different locations and collected after the completion of the tests.

The TDA material used in the experiments was obtained from a local tire-recycling centre in Montreal, Canada (Meguid and Youssef 2018). The material is categorized as Type-A with pieces that range in size from 30 to 120 mm. The properties of both the granular backfill and TDA material are summarized in Table 1.

Table 1. Properties of the backfill material

(a) Granular material[a]

Specific gravity	2.65
Coefficient of uniformity (C_u)	2.3
Coefficient of curvature (C_c)	1.6
Minimum dry unit weight (γ_{min})	15.1 kN/m^3
Maximum dry unit weight (γ_{max})	17.3 kN/m^3
Experimental unit weight (γ_d)	16.3 kN/m^3
Internal friction angle (ϕ)	47°
Cohesion (c)	0
Elastic Modulus (E)	150 MPa
Poisson's ration (v)	0.3

(b) Tire-derived aggregate

Maximum size	75 mm
Unit weight (γ)	6.2 kN/m^3
Uniformity coefficient (C_u)	1.9

[a]Meguid et al. (2017)

A total of six experiments were conducted including three benchmark tests where the instrumented pipe is surrounded by dry sandy gravel backfill only. For the remaining tests, a layer of TDA is placed above the pipe. For the benchmark tests, the placement of the backfill continued up to a distance 0.45 m above the crown. For experiments involving TDA, the granular backfill was used to make a 15 cm cover above the pipe, followed by the addition of TDA within a zone that measures 45 cm in width and 30 cm in height. The thickness of the aggregate cover placed above the pipe has been chosen to ensure that the protruding steel wires in the TDA (up to 10 cm in length) do not damage the pressure sensors. These dimensions were chosen to represent a compressible layer that has a thickness of about two times the pipe diameter. Two wood planks were used to assist in shaping the TDA zone while the rest of backfill material was added. The surface was finished by adding 10 cm of granular material over the entire backfill and levelled in the test chamber.

Surface load was applied using a loading platform placed under the actuator of the MTS machine. The load was gradually applied under displacement control with a displacement rate of 1.0 mm/min. The test would be stopped when either a surface displacement of 15 mm is reached or the pressures on the tactile sensors exceeded their allowable capacity (140 kPa). These limiting values were chosen in this study to avoid excessive pressure that could damage the tactile sensors. After the completion of each test, the tank was emptied using a vacuum machine connected to a collection barrel. The pipe was then retrieved and the setup was prepared for the next test.

3 Results and Discussions

The measured radial earth pressures are summarized in this section and the results are compared for the two investigated cases: (1) conventional backfill (benchmark tests); and (2) induced trench method using TDA inclusion. To validate the measured pressures in the benchmark tests, the sensor readings recorded following the placement of the granular backfill material are compared to Hoeg's analytical solution. The changes in earth pressure at selected locations on the circumference of the pipe are then presented.

3.1 Initial Condition

Hoeg (1968) developed an analytical solution for the contact pressure acting on buried cylinders. The solution is expressed in terms of two stiffness ratios (the compressibility ratio and the flexibility ratio) and three constants derived for two different boundary conditions, namely, no slippage or free slippage condition. The radial pressure (σ_r) is expressed as follows:

$$\sigma_r = \frac{1}{2}p\left\{(1+K)\left[1 - a_1\left(\frac{R}{r}\right)^2\right] - (1-K)\left[1 - 3a_2\left(\frac{R}{r}\right)^4 - 4a_3\left(\frac{R}{r}\right)^2\right]\cos 2\theta\right\}$$

(1)

where R is the pipe radius, r is the distance from the pipe center to the soil element under analysis, K is the lateral earth pressure coefficient expressed in terms of Poisson's ratio ($v/1 - v$), p is the soil vertical stress, θ is the angle of inclination from the springline and a_1, a_2 and a_3 are constants. The latter constants for the free slippage condition are given by:

$$a_1 = \frac{(1 - 2v)(C - 1)}{(1 - 2v)C + 1}$$

(2)

$$a_2 = \frac{2F + 1 - 2v}{2F + 5 - 6v}$$

(3)

$$a_3 = \frac{2F - 1}{2F + 5 - 6v} \tag{4}$$

The above constants are functions of the compressibility and flexibility ratios, which are measures of the extensional and flexural stiffnesses of the soil medium relative to the pipe. These rations are defined as follows:

$$C = \frac{1}{2} \frac{1}{1 - v} \frac{E_s}{\frac{E_p}{1 - v_p^2}} \left(\frac{D}{t}\right) \tag{5}$$

$$F = \frac{1}{4} \frac{1 - 2v}{1 - v} \frac{E_s}{\frac{E_p}{1 - v_p^2}} \left(\frac{D}{t}\right)^3 \tag{6}$$

where, E_s, v are the Elastic modulus and Poisson's ratio of the soil, respectively; E_p, v_p are the Elastic modulus and Poisson's ratio of the pipe, respectively; D is the average pipe diameter and t is the wall thickness. A soil-pipe system with both C and F of zero signifies a perfectly rigid embedded pipe.

The initial earth pressure calculated using Hoeg's analytical solution is demonstrated in Fig. 3. The pressure distribution is characterized by maximum values at the

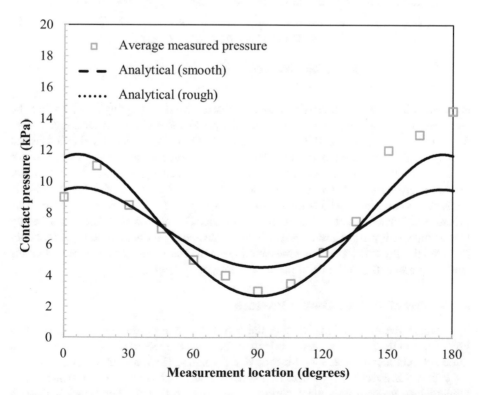

Fig. 3. Measured and calculated pressures on the pipe under initial loading condition

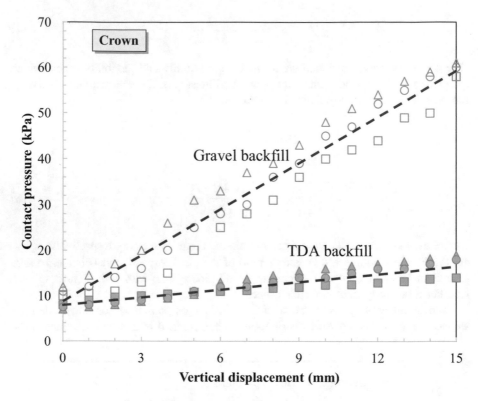

Fig. 4. Measured contact pressures at the crown

crown and invert which quasi-sinusoidal decrease towards the springline (SL). For the free and no slippage boundary conditions, the pressures at the crown and invert were found to be about 12 kPa and 9 kPa, respectively. These values decreased to 3 kPa and 5 kPa, respectively, at the springline. By superimposing the measured pressure values on the analytical solution, it is evident that the experimental results followed a similar pressure distribution pattern with an increase of about 2 kPa at the invert. This is consistent with additional in-situ pressure at a depth of $1D$ below the crown. This pressure difference is not usually captured by Hoeg's solution that assumes a deeply buried pipe with vertical pressure that is the same above and below the pipe. It should be noted that the contact pressure measured using the tactile sensors and reported in this study represents the radial earth pressure acting on the pipe.

3.2   Effect of TDA on Contact Pressure

To evaluate the effect of introducing the soft TDA zone on the distribution of earth pressure onto the pipe, a comparison is made between the recorded pressure readings at three selected locations, namely: crown (C), springline (SL), and invert (I).

Figure 4 illustrates the changes in contact pressure at the crown with the increase in applied load up to a maximum surface movement of 15 mm. The benchmark tests

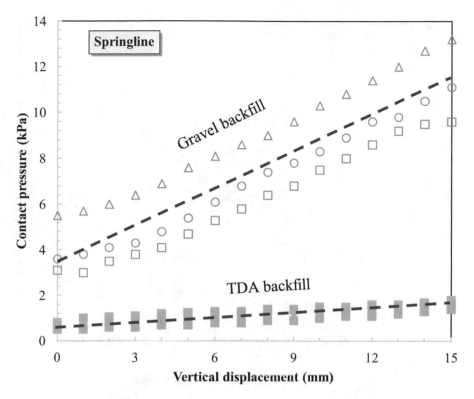

Fig. 5. Measured contact pressures at the springline

showed consistent increase in radial pressure from the initial condition (about 10 kPa) up to a maximum value of about 60 kPa. Tests involving the TDA inclusion above the pipe resulted in a significant decrease in radial pressure at the crown from 60 kPa to as low as 18 kPa. This is attributed to the soil arching effect and load re-distribution around the pipe; ultimately a consequence of the soft-zone presence.

The earth pressure at the springline increased from 4 kPa to as high as 11 kPa in the benchmark tests as shown in Fig. 5. Contact pressure decreased from 11 kPa in the benchmark tests, at applied displacement of 15 mm, to less than 2 kPa, using the induced trench method with TDA inclusion.

The highest measured pressure was recorded at the invert of the pipe where pressure reached about 70 kPa when gravel backfill was used (see Fig. 6). This value decreased significantly to less than 20 kPa using TDA inclusion. The pressure distribution on the pipe at all investigated locations is summarized in Fig. 7. The contact pressure is normalized with respect to the benchmarks tests. It is found that contact pressures decreased at all locations as the surface displacement increased from 0 mm (initial condition) to 15 mm. The pressure ratio at the beginning of the test was found to be 70%, 52% and 20% at the crown, invert and springline, respectively. As the load is applied and the induced trench took effect, the pressure ratio changed to 30%, 25% and 13% at the crown, invert and springline, respectively when the applied displacement

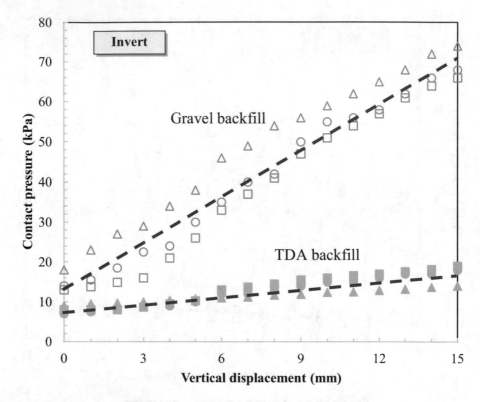

Fig. 6. Measured contact pressures at the invert

reached 15 mm. This suggests that utilizing TDA above buried pipes can result in a significant reduction in earth load and consequently, an economic design.

Based on the above results, the effect of TDA in reducing radial earth pressure on the pipe is evident. In addition, the effect seems to become more pronounced at high surface loads where sufficient compression develops in the TDA zone and consequently more shear stresses are generated at the boundaries of the induced trench. Given the increasing amount of waste materials that may be disposed in landfills around the world, re-using scrap tires in buried structures seems to be a sound alternative both technically and environmentally.

4   Summary and Conclusions

This study investigated the earth pressure distribution on a buried pipe installed using the induced trench method. The contact pressure distribution on the pipe was measured using the tactile sensing technology, which is able to provide a continuous pressure profile on the pipe wall. The effect of installing a soft zone of TDA on the radial pressure distribution on the pipe was examined.

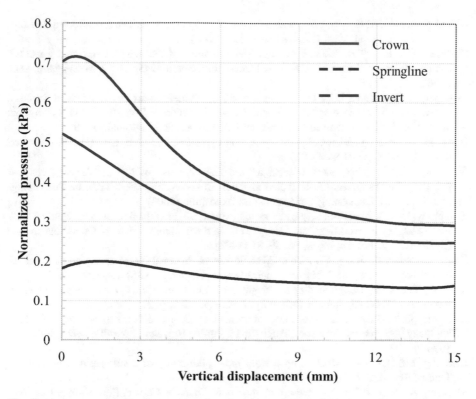

Fig. 7. Average normalized pressures for the case of TDA inclusion with respect to the measured values for gravel backfill

The induced trench installation described in this study was successful in reducing the vertical loads on the buried pipe. The average measured earth pressure above the crown of the pipe was found to be about 30% of the overburden pressure for installations with granular backfill material. Significant reduction in radial pressure was also recorded at the invert with pressure reduction of about 77% with the introduction of the soft TDA inclusion.

This study suggests that using TDA in induced trench construction is both technically and environmentally sound alternative to gravel backfill and EPS geofoam.

Acknowledgements. This research is supported by a research grant from the Natural Sciences and Engineering Research Council of Canada (NSERC). The authors would like to acknowledge the support of Dr. W.D. Cook and Mr. J. Bartczak in conducting these experiments.

References

Ahmed, M., Tran, V., Meguid, M.A.: On the role of geogrid reinforcement in reducing earth pressures on buried pipes. Soils Found. **55**(3), 588–599 (2015)

ASTM Standard D6270: Standard practice for use of scrap tires in civil engineering applications. ASTM International, West Conshohocken, PA (2008)

Bosscher, P.J., Edil, T.B., Eldin, N.N.: Construction and performance of a shredded waste tire test embankment. Transp. Res. Rec. J. Transp. Res. Board **1345**, 44–52, Washington, D.C. (1992)

Drescher, A., Newcomb, D.E.: Development of design guidelines for use of shredded tires as a lightweight fill in road subgrade and retaining walls. Report No. MN/RC-94/04, Department of Civil and Mineral Engineering, University of Minnesota, Minneapolis (1994)

Hoeg, K.: Stresses against underground structural cylinders. Am. Soc. Civ. Eng. Proc. J. Soil Mech. Found. Div. **94**(SM4), 833–858 (1968)

Humphrey, D.N.: Effectiveness of design guidelines for use of tire derived aggregate as lightweight embankment fill. Recycled materials in geotechnics (GSP 127). In: Proceedings of ASCE Civil Engineering Conference and Exposition (2004)

Humphrey, D.N.: Tire derived aggregates as lightweight fill for embankments and retaining wall. In: International Workshop on 'Scrap Tire Derived Geomaterials – Opportunities and Challenges', Yokosuka, Japan, pp. 59–81 (2008)

Kang, J., Parker, F., Kang, Y.J., Yoo, C.H.: Effects of frictional forces acting on sidewalls of buried box culverts. Int. J. Numer. Anal. Methods Geomech. **32**(3), 289–306 (2008)

Kim, K., Yoo, C.H.: Design loading for deeply buried box culverts. Highway Research Center. Report No. IR-02-03. Auburn University, Alabama, USA, p. 215 (2002)

Leonards, G.A., Roy, M.B.: Predicting performance of pipe culverts buried in soil. Joint Highway Research Project Report JHRP-76-15. Purdue University, West Lafayette, Ind., May 1976

Liedberg, N.S.D.: Load reduction on a rigid pipe: pilot study of a soft cushion installation. Transp. Res. Rec. **1594**, 217–223 (1997)

Marston, A.: Second Progress Report to the Joint Concrete Culvert Pipe Committee. Iowa Engineering Experimental Station, Ames (1922)

Marston, A.: The theory of external loads on closed conduits in the light of the latest experiments. In: The 9th Annual Meeting of the Highway Research Board. Highway Research Board, Washington, D.C., USA, pp. 138–170 (1930)

McGuigan, B.L., Valsangkar, A.J.: Centrifuge testing and numerical analysis of box culverts installed in induced trenches. Can. Geotech. J. **47**(2), 147–163 (2010)

Meguid, M.A., Hussein, M.G., Ahmed, M.A., Omeman, Z., Whalen, J.: Investigation of soil-geosynthetic-structure interaction associated with induced trench installation. Geotext. Geomembr. (2017a) https://doi.org/10.1016/j.geotexmem.2017.04.004

Meguid, M.A., Ahmed, M.R., Hussein M.G., Omeman, Z.: Earth pressure distribution on a rigid box covered with U-shaped geofoam wrap. Int. J. Geosynth. Ground Eng. **3**(11) (2017b). https://doi.org/10.1007/s40891-017-0088-4

Meguid, M.A., Hussein, M.G.: A numerical procedure for the assessment of contact pressures on buried structures overlain by EPS geofoam inclusion. Int. J. Geosynth. Ground Eng. **3**(2), 2 (2017c). https://doi.org/10.1007/s40891-016-0078-y

Meguid, M.A., Youssef, T.: Experimental investigation of the earth pressure distribution on buried pipes backfilled with tire-derived aggregate. Transp. Geotech. **14**(1), 117–125 (2018)

Mills, B., McGinn, J.: Design, construction, and performance of highway embankment failure repair with tire-derived aggregate. Transp. Res. Rec. J. Transp. Res. Board **1345**, 90–99, Washington, DC (2010)

Oshati, O.S., Valsangkar, A.J., Schriver, A.B.: Earth pressures exerted on an induced trench cast-in-place double-cell rectangular box culvert. Can. Geotech. J. **49**(11), 1267–1284 (2012)

Paikowsky, S.G., Palmer, C.J., Dimillio, A.: Visual observation and measurement of aerial stress distribution under a rigid strip footing. In: Geotechnical Special Publication, vol. 94, pp. 148–169 (2000)

Sladen, J.A., Oswell, J.M.: The induced trench method—a critical review and case history. Can. Geotech. J. **25**(3), 541–549 (1988)

Springman, S.M., Nater, P., Chikatamarla, R., Laue, J.: Use of flexible tactile pressure sensors in geotechnical centrifug. In: Proceedings of the International Conference of Physical Modeling in Geotechnical Engineering, pp. 113–118 (2002)

Strenk, P.M., Wartman, J., Grubb, D.G., Humphrey, D.N., Natale, M.F.: Variability and scale-dependency of tire derived aggregate. J. Mater. Civ. Eng. **19**(3), 233–241 (2007)

Sun, L., Hopkins, T., Beckham, T.: Long-term monitoring of culvert load reduction using an imperfect ditch backfilled with geofoam. Transp. Res. Rec. **2212**, 56–64 (2011)

Tweedie, J.J., Humphrey, D.N., Sandford, T.C.: Full scale field trials of tire shreds as lightweight retaining wall backfill, at-rest condition. Transportation Research Board, Washington, DC (1998)

Vaslestad, J., Johansen, T.H., Holm, W.: Load reduction rigid culverts beneath high fills: long-term behavior. Transp. Res. Rec. **1415**, 58–68 (1993)

Xiao, M., Ledezma, M., Hartman, C.: Shear resistance of tire-derived aggregate using large-scale direct shear tests. J. Mater. Civil Eng. **27**(1), 1–8 (2015)

Application of Electroosmosis and Monitoring for the Management of Geotechnical Processes in Underground Construction

Nicolay Perminov[✉]

Docent, Emperor Alexander I St. Petersburg State Transport University,
St. Petersburg, Russia
perminov-n@mail.ru

Abstract. A new concept for the selection of rational construction and engineering parameters for the building of large deep-set edifices is set out for an environment where the requirements on preservation of historical sites and rational usage of land in big cities are growing more taxing. Based on the proposed geotechnical model, and the assessment of the results of numerical modeling, research and experiments, a number of techniques have been developed intended to optimize the technological modes of construction under adverse urban conditions involving hard-to-handle soils and congested urban development.

1   Introduction

Long-term experience of construction, rehabilitation and exploitation of a megalopolis underground engineering infrastructure has been used to develop a concept and philosophy of establishing and operating the safety geomonitoring system.

The article describes the philosophy of building the space-time structure of the geomonitoring: programs for calculating and forecasting the changes of geotechnical conditions and geomassive deflected mode under different building erection regimes with the use of electro-osmosis; the system of technical facilities used for instrumental observations and for checking the change of specific elements within the "structure-geomassive" system; the data measurement system to collect, process, store and identify the observation vibration control parameters; integrated geotechnical methods of direct impact on the soil bulk and protection of underground engineering infrastructure.

The article presents the experience of monitoring and taking protective actions, aimed to ensure safety underground engineering infrastructure at the depth of 70 m, located in the zone affected by geotechnical influence of the Underground Infrastructure and high-rise construction. A comparative analysis of calculated-experimental data and the results of the many-year monitoring aimed to ensure safety of the underground wastewater collection and treatment facilities in St. Petersburg, is given.

© Springer Nature Switzerland AG 2019
M. Badr and A. Lotfy (Eds.): GeoMEast 2018, SUCI, pp. 94–104, 2019.
https://doi.org/10.1007/978-3-030-01884-9_8

2 Application of Geotechnics and Monitoring in Solving Problems of Construction of Unique Underground Structures in Difficult Conditions

Today, the problem of working out the complex measures providing high quality building turns out to be very relevant. It could be explained by growing demands towards ecology, earth resources and environment protection against negative technogenic influence Geo monitoring systems rank among such measures.

The experience in design and construction of sewage treatment systems and constructions has proved the necessity of such systems. It is well-known that for such purposes main pump stations are usually built using dipping method.

Today St. Petersburg and many large cities all over the world are facing the problem of metro construction connected with the lack of free territory. In this connection we propose to build junction metro stations of St. Petersburg metro in large-sized dipping wells (Fig. 1).

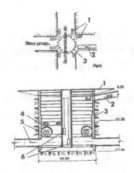

Fig. 1. Multifunctional deep-set junction terminal (Location), (Junction terminal plan) 1-Pedestrian cross; 2- Underground garage entrance; 3- Deep-set edifice 4- Underground garage; 5-Junction terminal stations; 6- Lift shaft

We propose to use a construction of 66.0 m in diameter and 70.0 m high for that purpose. It will allow to organize the junction terminal in the lower part of the well. The upper part could be used for a 7-storied underground garage for 850 cars. Due to that we would save 15000 m^2 of the city territory. Usually the constructions of the described type are deepened into the ground for 70 m. Their cross section ranges from 2000 to 3000 m^2. So the contacting area between their lateral surface and the ground is from 150 to 200 thousands of sq. meters. When these constructions cross several aquifer levels (above 5) and greatly influence upon the surrounding buildings and the geological environment.

The type and the after-effect of such influence differ from time to time. On one hand, they are determined by the engineering and geological conditions of the construction site. On the other hand, - by the designing and technological peculiarities of the deepened construction and its building. Hence there is a problem how to exclude or minimize the above mentioned negative after-effects. To be settled this problem needs the systematic approach towards joint solving geotechnical and engineering problems.

Today we are undertaking the works upon geo- monitoring system creation. This system is being implemented step by step in the Northern and Southern sewage systems construction.

The geo-monitoring system structure is based on the following subsystems:

- the program system calculating and forecasting the changes in technical and geological conditions and strain-stress state (SSS) of a geological block under different building modes for dipping constructions;
- the technical tools subsystem for instrument observations and monitoring of "construction - geo block" system elements;
- informational and measuring subsystem for gathering, processing storing and identification of observation and monitoring data;
- the system of geo-technological methods for determined operation upon a soil bulk and construction;
- automation and optimum control subsystem for geo-technological and other processes.

We have large experience in implementation of different methods providing reliability and quality in deepening constructions building. Our experience shows that geo-monitoring system should be adapted to control the processes in semi-continuous mode and in real time.

Some research institutes used strained and deformed state control system for deep wells construction. The results show that every deviation from the project is registered by the control instruments.

However it's a great problem to use these data for determined process control. It is explained by the fact that it takes much time from getting the initial data to the moment of its analysis and issuing the recommendations for some technological steps. It is connected with the long period of time needed for data processing. Unfortunately the structural and geo-engineering conditions also change meanwhile.

Our investigations have shown that it's possible to provide the reliability and uniformity for edifices deepening process. It could be achieved by creating the semi-continuous interaction system for the three- dimensional (dX, dY. d Z) and time (dt) process variables.

We have taken out some estimate-theoretical and research works. As a result we have worked out and tested the complex system providing the geotechnical support for deep-set edifices deepening process. These edifices were from 50 to 70 m in diameter. They had been implemented in waste treatment facilities construction in St. Petersburg.

The mentioned above system comprises three following complexes: measuring-control complex; signal estimation and transmission for process control purposes; immediate feed-back complex.

The measuring-control complex provides monitoring for the SSS orientation of edifice shell and soil bulk. It includes spatial orientation instruments (laser range finder, volumetric reflectors, bank angle detectors). Besides, it includes the instruments estimating ground resistance along the lateral surface of the building, ground lateral pressure, ground pressure under the edifice knife banquet, reinforcement stress, concrete distortion.

SSS control system provides monitoring of the following variables characteristic for the "deep-set edifice - geomass" behaviour:

1. Ground SSS, that is estimated by the results of contact pressure, pore pressure and ground shift detection;
2. Edifice material SSS concrete and reinforcement including;
3. Deep set edifice caisson and banks.
4. Ground subsidence (on the adjacent territory).

Initial transformers are installed in 6–8 section points in edifice shell perimeter from the 5th to 10th layers.

E.g.: There were 86 pressure, friction temperature and other transformers installed in the deep-set edifice of 49.6 m diameter.

Strain sensors, steam pressure detectors, string hydraulic sensors and rigid string hydraulic sensors were used for contact pressure measuring.

The results of reinforcement strain and distortion detection were used to estimate the Edifice shell SSS. String detectors and strain sensors that were stacked to the reinforcement were used for that purpose.

Edifice bank was detected by the bank sensors. Those sensors are based on double string gauging transformers (measuring accuracy didn't exceed 10 s).

Semi-continuous system for detectors indication registration includes the following instruments: portable digital lime setter, 20 channel electronic communicator, controller (stand alone interface module) Kl-20; controller remote control unit; power supply unit; wiring cables (controller, controller - communicator, power supply - controller; PC controller, power supply - battery, communicator, periodic tolling controller, communicator - sensors).

We have proposed new geotechnical methods for unique large-scale edifices. These methods are based on geo-mass SSS controlling not only in the basement but also in its lateral surface.

Essentially the described methods and their realization technique are the following:

- Geo-mass natural SSS estimation is fulfilled at the designing stage through numerical calculations (e.g.: Finite elements method).
- Geo-mass SSS and well orientation forecast are worked out on the basis of the above mentioned calculations. Besides we can build a model of edifice vertical angular deviation.
- We can forecast the location of deepening marks with the damage emergency. On the basis of such forecasts we can select the most acceptable geo- technical methods for geo-mass SSS measuring not only at the edifice base but also in the lateral surface. Besides we can select the methods to change the conditions for their contact interaction.
- We can make the technical-economic estimation and select the optimum method to provide deepening process reliability and acceleration.

We had fulfilled several cycle calculations with Finite elements method for the soil conditions typical for St. Petersburg (Quaternary sediments 26–35 m deep, underlain by hard Cambrian clays).

As a result of computing experiment we had calculated the admissible deviation 'd_h' from the vertical axes and non-deforming fits 'd_i' values. We have come to a conclusion that tixothropic jacket is the best possible solution in this case. It results in friction strength decrease on the marks from 16 to 47 m. Hovering stability is achieved through electroosmosis up to 16 m height and with 'd_i' < 0.015. We propose the dedicated method for geomass preparation to correct the bank for 'd_h' and d_i high values. Besides, what we propose is the discharge well design in knife and lateral surface areas.

The described above results using geomonitoring system for large diameter deep-set edifices construction could be successfully implemented for large scale basements construction. Besides it might be used for deep-set edifices construction in hard and complicated ground conditions. Also these methods could be used for developing and reconstruction of the existing underground territories in large cities.

3 Structural and Geo-Technological Model for Large Scale Deep-Set Edifices Construction Under City Building System Conditions

To save city building environment we have worked out the structural geo-technological model of large scale edifice construction. This model is created considering the characteristic features of interaction between structural and technological processes, on one hand, and engineering and geological processes, on the other.

The monitoring object - the deep-set edifice is represented as a geotechnical system (GTS) - that is the interconnected and interlined assembly of technical object (TO) (the deep-set construction itself) and of geological object (GO) (the ground).

Technical object is a spatial one with the concentrated influence and distributed structural and technological parameters. Geologic object is an assembly of city building and geomechanical elements that might be combined corresponding to geological environment model (Fig. 2).

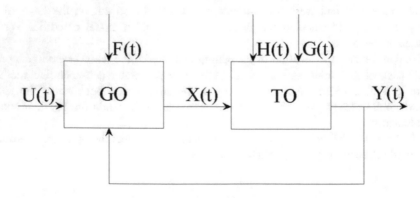

Fig. 2. Geotechnical system

GTS is influenced by a great number of factors that could be divided into three following groups:

- construction elements mass (G);
- geological environment reaction (X);
- and external (artificial and natural) disturbance (H).

Each vector: G, H, X, consists of numerous factors. The greater part of those factors is unknown. To determine them multidimensional distribution law for probability distribution should be defined.

Please, find bellow the equation that describes the process of deep-set edifice construction and functioning. This equation describes it in a simplified mode according to the common principals.

$$a(H, G, X, t)d\ddot{I}/dt = R - Re(x) \tag{1}$$

$$R = Re + Rp \tag{2}$$

where $\ddot{I}$ is a reduced factor of target product output (by the specific volume of deep-set edifice internal volume),

R, Re, Rp - reduced expenditures (reduced costs, power and labour expenditures); t - time of functioning.

The task for optimum GTS monitoring could be formulated in the following way: one is to find the control rules U(t) for GTS. taking into consideration a set of regulations. These regulations transfer technological and geological objects to the monitoring mode. Optimum function should have an extreme and is expressed with the following equation:

$$Fn = \int_{t0}^{tn} P(Y, U, II, G, t)dt \ max \tag{3}$$

where P is continuous function of output variable Y and of controlling influences H, G, U.

Optimum criteria for the whole period of GTS development and functioning (e.g.: deep-set well construction) is represented as a sum optimum criteria elements on each construction stage (installation and deepening).

$$F = \sum Fn \tag{4}$$

Optimum GTS control is reduced to the step by stem TO and GO variables control. The control is fulfilled according to the algorithm of optimum way of aim achievement. So our aim is to settle the problem of deepening the well shell under the optimal geo-technological conditions in the preset time or with the minimum influence upon city building environment.

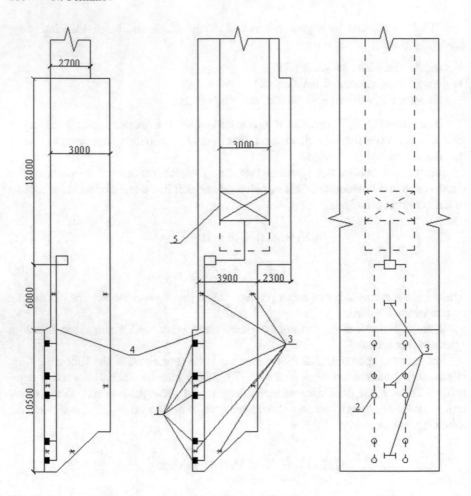

Fig. 3. Instruments location plan on the deep- set well: 1,2 - pressure and friction sensors; 3 - reinforcement dynamometer; 4 - wiring cables; 5 - control station.

4   Geotechnical Research and Manufacturing Results

Geo-technical research works include large-scale laboratory experiments, numerical modeling by Finite elements method and in situ tests.

To simulate the conditions of deepening by Finite elements method we used incremental soil model, based on generalized Gook principle. The relation between strain and deformation increments was computed separately for stress tensor deviation and spherical components:

$$dS_{ij} = 2G^T * de_{ij} \tag{5}$$

$$d\sigma_{cp} = 3K^T * d\varepsilon_{cp} \tag{6}$$

where dS_{ij} and $d\varepsilon_{ij}$ - stress and deformations tensors deviation components increments respectively;

$d\sigma_{cp}$ and $d\varepsilon_{cp}$ - average stress and deformation values increments;

G^T and K^T - tangent modules of shape and object deformation.

First in laboratory and then in situ we have carried out the investigation to find out the tixothropic solution features influence upon the large scale deep-set edifices deepening process. The plan of monitoring instruments location is shown in Fig. 3.

The results of friction and pressure strength detection are shown in Fig. 4. Engineering and geological conditions of the construction site are typical for St. Petersburg city building environment. It means that the territory is formed by the grounds of Quaternary sediments 26–35 m deep, underlain by semi-hard and hard Cambrian clays up to the 71.0 m depth.

The well for Northern waste treatment facilities with 66.1 m diameter is designed as a monolith unit. We had included a tixothropic jacket into construction to provide the deepening process reliability. Deepening process numerical modeling for large scale wells in the inclined layers of Cambrian clays had shown the necessity of control system for deepening process. The system included electroosmosis plant and instrumental control net.

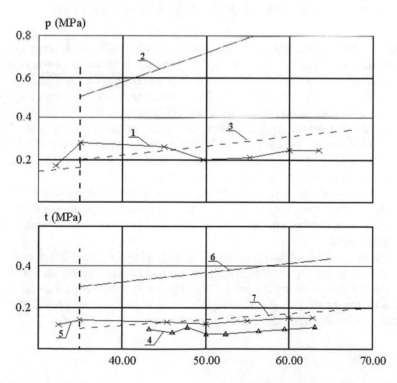

Fig. 4. Pressure and friction measurements during the deepening process. 1, 4, 5 - experimental data; 3, 6, 7 - computering values for ground pressure and friction.

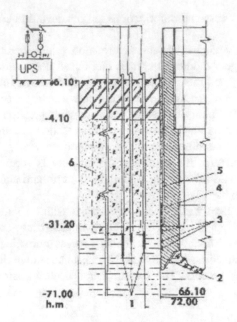

Fig. 5. Scheme for electroosmosis edifice deepening. 1,2 - tube electrodes and electrode belt; 3-control instuments; 4- clay solution; 5 tixothropic solution;6 - ice ground wall.

Electroosmosis plant comprised the electrode bell 10.5 m high mounted on the knife external surface and 45 tube electrodes. The electrodes were located along three concentric circles. The distance from the well external ring was 3, 6 and 7 m. They were dipped to the 41.0–43.0 m height. The rectifiers were used as power supply units for DC voltage. They had provided 74 V and 3300 A (Fig. 5).

The controlled deepening process was conducted in the following way. First the ground was electrically treated with for 1.5–3.0 h (voltage gradient was 0.2 V per sm, current density was about 2 A/sq m). Then the well was gradually deepened. Previously the well had been motionless for several days (Fig. 6).

Preliminary electrical treatment allowed to increase the deepening rate 2.5–3.0 times. The deepening rate ranged from 5 sm to 15 sm per minute, the electrode polarity being changed deepening rate abruptly went down, the bank was corrected and the deepening stopped.

For Southern waste treatment facilities the deepening well was 51.0 m in diameter. The deepening height was 49.0 m. Anodes and cathodes were located on the knife part of the construction in scattering order. The clay solution had the access to the electrodes. As a result we achieved deepening acceleration, bank correction and eliminating of negative influence upon city building environment.

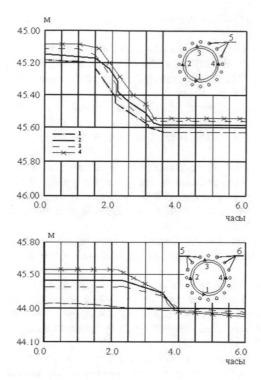

Fig. 6. Electroosmosis deepening diagram phragments a - even deepening; b - bank correction; 1,2,3,4 - fixed points 5,6 - electrodes (catodes and anodes)

5 Conclusions

Concept of the geomonitoring ensuring of technosphere safety of construction and maintenance of underground infrastructure, principles of its formation and functioning should provide reliability, safety and efficiency of underground structures construction. In some cases they can be successfully used at engineering development of Urban Underground Infrastructure space of a megalopolis, when it is necessary to provide protection of town-planning environment against negative technical impact. In other cases, vice versa, they can be used to provide preservation of underground spot or line structure being erected against change in condition of exposure of town-planning environment or other poly-anthropogenic factors, which disturb the conditions of technosphere safety.

References

Bridge and culvert limit pase on big earth and rock project, Highway and Heavy Constr. **129**(7), 40–41 (1986)

Grammond, N.Y.: Numerical X-ray diffraction analysis of ettringite, thaumasite and gypsum in concrete and mortars. Cem. Concr. Res. **15**(3), 431–441 (1985)

Hebden R.H.: Segmental precast culverts for the coquihalla Fraway Concr.Transp.Detroit 769–795 (1986)

Hobbs D.W.: Expansion and cracking attributed to delayed ettringite formation. In: Proceedings of a Technical Session Ettringite, pp. 151–181. ACI. Seattle, Washington, SP-177 (1999)

Ivliev E.A. (2015), Soil mechanics and foundation engineering, Grounds, foundations and mechanics of soils, # 6. pp. 14–18

Kolos A., Konon A.: Estimation of railway ballast and subballast bearing capacity in terms of 300 kN axle load train operation. In: Zhussupbekov, A. (ed.). Proceedings of International Conference Challenges and Innovations in Geotechnics. Balkema, Rotterdam, 5–7 August (2016)

Pankova G.A., Klementiev M.N.: The experience of exploitation of sewage tunnels in St. Petersburg, Water supply and sanitary equipment, # 3. pp. 55–61 (2015)

Perminov N.A.: Complete geotechnical and monitoring services for the construction of the underground structure in a megapolis. In: Proceedings of the International Geotechnical Conference, pp. 361–366. Almaty (2004)

Perminov N.A.: Geotechnical aspect of safety assurance for long-used engineering infrastructure facilities in large cities in complicated ground conditions. Geotech. Roads Railw. 1195–1201 (2014)

Pinto A., Tomasio R., Pita X., Pereira A., Peixoto A. (2011), Gutter soil mixing solutions in Portugal on hard soil and weak rocks. In: Proceeding of the 15th European Conference On Soil Mechanics and Geotechnical Engineering, vol. 2, pp. 1037–1042. Athina

Recommendation for Design and Analysis of Earth Structures using Geosynthetic Reinforcements –EBGEO, 2nd edn. (2011)

Serebryakov D.V.: Of the study on vibrational process of soil roadbed on the sections of the pairing of railway track with bridges. In: Proceedings of International Scientific Conference Transportation System Infrastructure Problems, vol. 240, pp. 52–55 (2015). Saint-Petersburg, Russia

The Main Directions of Providing Reliability of Exploitation of Railway Engineering Structures at the Contemporary Technical and Technological Level (Resolution of 14.08.2013 # 275)

The Technical Conditions for Conducting Planned and Preventive Repairs of Railway Engineering Structures in Russia for the Central Track Service -622 (2008)

Tikhomolova, K.P.: Electroosmosis, p. 248. Khimiya, Leningrad (1989)

Vasiliev V.M., Pankova G.A., Stolbikhin Yu.V.: Destruction of sewage tunnel and above structures due to the impact of microbiological corrosion, Water supply and sanitary equipment, # 9, pp. 55–61 (2013)

Wells, T., Melchers, R.E.: Modelling concrete deterioration in sewers using theory and field observations. Cem. Concr. Res. **77**, 82–96 (2015)

Ground Response and Support Measures for Rohtang Tunnel in the Himalayas

L. M. Singh[1], T. Singh[2(✉)], and K. S. Rao[2]

[1] Rohtang Tunnel, Manali, Himachal Pradesh 931720, India
[2] Department of Civil Engineering,
IIT Delhi, Hauz Khas, New Delhi 110016, India
`tarun.iitd@outlook.com`

Abstract. The Rohtang tunnel is about 8.8 km long and is one of the important road tunnels in the Himalayas which facilitates all round connectivity between Manali (in the south) and Keylong (in the north), with its Lahul & Spiti district headquarter. This tunnel lies in complex geology with many folds and faults intersecting it at various locations. Three major rock types include uniformly dipping sequences of phyllites, schists and gneissose rocks with a minimum overburden of about 600 m and a maximum of 1900 m. The geological complexity of area further increases due to anatexis and migmatization of rocks. In order to assess the tunnel stresses and displacement, a closed-form solution is used. The obtained results are validated using finite element analysis. The role of complex geology and high & varying overburden is considered for suitable designing of tunnel support system.

Keywords: Rohtang tunnel · FEM · Closed-form solution
Numerical analysis

1 Introduction

A great number of transport and hydro tunnels are under construction in highly tectonic and young Himalayan Mountains. Being tectonically active these mountains consist of weak, heterogeneous, deformable, fractured rock mass with high degree of weathering and ingress of water. Squeezing of tunnel is quite common in weak, anisotropic rocks such as phyllites, schist, quartzite, migmatite and in shear/fault zones. The degree of squeezing is further enhancing if tunnels passing through these rocks mass contain higher overburden. Therefore, during construction, the stability of tunnel is of major concern in such areas. Though swelling, rock bursting is occasional but squeezing is a very common phenomenon in Himalayas leading to collapse of tunnel. The 8.8 km long Rohtang tunnel encompasses various formations from South to North Portal. This study has been done to assess ground response of Rohtang tunnel with the help of finite element analysis and closed form solution. The main emphasis has been given in shear zone in the tunnel sections and suggested stability measures in the same tunnel sections.

© Springer Nature Switzerland AG 2019
M. Badr and A. Lotfy (Eds.): GeoMEast 2018, SUCI, pp. 105–125, 2019.
https://doi.org/10.1007/978-3-030-01884-9_9

2   Factors Influencing Tunneling

As the rock mass is having heterogeneity so various factors which affect the rock mass. Mainly two factors rock mass quality and mechanical processes on which stability of underground structure like tunnels, caverns depends. The stability of these structure also depends on shape, size and orientation of the structure. Figure 1 shows the factors influencing the stability of underground structure.

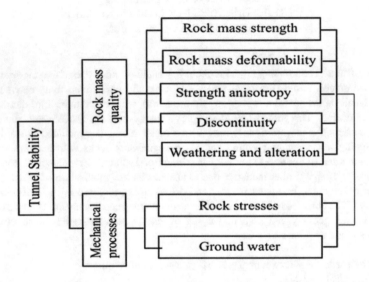

Fig. 1. Factors influencing on tunnel stability

2.1   Rock Mass Quality

The rock mass quality mainly depends on strength of rock, deformability of rock, anisotropy, discontinuity, and weathering and alterations of rock masses.

Rock Mass Strength
Due to presence of discontinuities, rock mass is having very low strength so it takes very low stress and deformation comparison to the intact rock. It is very difficult to determine the strength of rock mass in laboratory as it requires undisturbed and truly representative sample with discontinuities. So with the help of intact rock properties which is to be measured in the laboratory with other properties many authors proposed empirical formulas to estimate rock mass strength. Mostly used empirical relations As per Rock Mass Rating (RMR) (Bieniawaski 1989, 1993), Hoek-Brown relationship (Hoek et al. 2002) and Q-value correlation (Barton 2002) given in Table 1.

Where; rock mass unconfined compressive strength σ_{cm} in MPa, intact rock uniaxial compressive strength σ_{ci} in MPa, RMR is the Bieniawaski's rock mass rating, s and a are the material constant related to Hoek-Brown failure criteria GSI is the geological strength index, γ is the rock density in t/m^3 and Q is the rock mass quality rating.

Table 1. Empirical relations of rock mass strength.

Proposed by	Rock mass strength and its relationship with rock mass classifications
Bieniawaski (1993)	$\sigma_{cm} = \sigma_{ci} \times exp\left[\frac{RMR-100}{18.75}\right]$
Hoek et al. (2002) and Hoek (1994)	$\sigma_{cm} = \sigma_{ci} \times \left[exp\left(\frac{GSI-100}{9}\right)\right]^a = \sigma_{ci} \times \left[exp\left(\frac{RMR-105}{9}\right)\right]^a$
Barton (2002)	$\sigma_{cm} = 5\gamma \times \left[\frac{\sigma_{ci}}{100} \times Q\right]^{\frac{1}{3}}$

Rock Mass Deformability

Due to presence of discontinuities the deformability of rock mass increased in comparison to intact rock. The deformation modulus is an important property to design of underground structure (Dere et al. 1967; Dershowitz et al. 1979; Bieniawski 1978). The ISRM (1975) defines the deformation modulus is the ratio of stress and corresponding strain at the time of loading.

There are so many methods like Flat Jack Test, Plate Jacking Test, Cable Jack Test, Goodman Jack Test, Plate load test etc. are available to determine the deformation modulus in the field. The methods mentioned above are time consuming, high cost oriented and difficult in execution in such type of terrain. The result obtained are differ considerably (Nilsen and Palmstrøm 2000). To estimate the rock mass deformation many empirical formulas are given by different researchers given in Table 2.

Table 2. Empirical relation of rock mass deformation modulus.

Proposed by	Rock mass deformation modulus
Hoek and Brown (1997)	$E_m = \sqrt{\frac{\sigma_{ci}}{100}} \times 10^{\left(\frac{GSI-10}{40}\right)}$
Serafim and Pereira (1983)	$E_m = 10^{\left(\frac{RMR-10}{40}\right)}$
Palmstrøm (1995)	$E_m = 5.6 \times RMi^{0.375}$
Bieniawaski (1978)	$E_m = 2RMR - 100$
Barton (2002)	$E_m = 10 \times \left(\frac{Q \times \sigma_{ci}}{100}\right)^{\frac{1}{3}}$

Where; E_m is the rock mass deformation modulus in GPa and RMi is the Palmstrom's rock mass index.

Strength Anisotropy

Due to tectonic forces the rocks in Himalayas are anisotropic having strength and deformation in multidirectional. So many times due to slabbing and thin bands of schist, phyllite which encounter within the strong rock such as quartzite, gneiss and dolomite which is having altogether different rock mass characteristics. As the rock mass properties are weaker, high schistocity and low friction are having very low self-supporting capacity facing severe stability problems during construction of tunnel.

Many research has been carried out in the Himalayas as well as other part of the world which shows that the uniaxial compressive strength of intact rock will be minimum at 30 degree with the loading direction ($\beta = 30$ degrees), and maximum when this schistocity plane will be at 90 degree from the loading direction ($\beta = 90$ degrees). Figure 2, illustrates the phenomenon of above on the results of different researches.

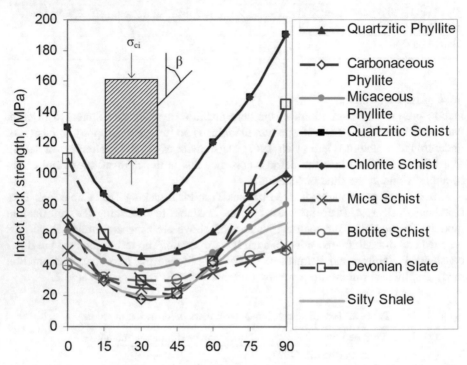

Fig. 2. Uniaxial compressive strength at different angle of schistocity plane (after Ramamurthy 1993; Hawkins 1998; Ajalloeian and Laskaripour 2000 and Nasseri et al. 2003).

Discontinuity
The discontinuity is nothing but separation of rock continuum having zero tensile strength. It is represented by spacing, frequency, orientation, persistence, roughness, alteration, weathering, aperture, block size (Barton et al. 1985; Hudson 1989). As per ISRM (1978) the parameters which are to be considered to define the rock mass discontinuity as illustrated in Fig. 3.

Rock Weathering
Weathering of rock is a natural phenomenon which changes the rock mass property even of the same composition. The effect of weathering will be maximum at the surface and decreases as the depth increases as illustrated in Fig. 4. Weathering reduces the strength of rock, slake durability, resistance to friction and same time it increases porosity, permeability. Beavis et al. (1982) and Gupta and Rao (2000) evaluated the weathering effect on the rock mass properties such as porosity, density, tensile strength, uniaxial compressive strength and elasticity modulus.

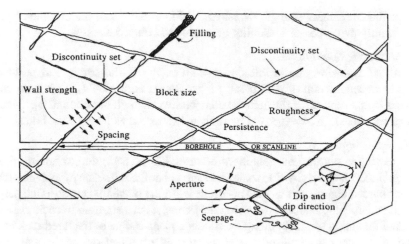

Fig. 3. Discontinuity characteristics in the rock mass (Hudson and Harrison 1997).

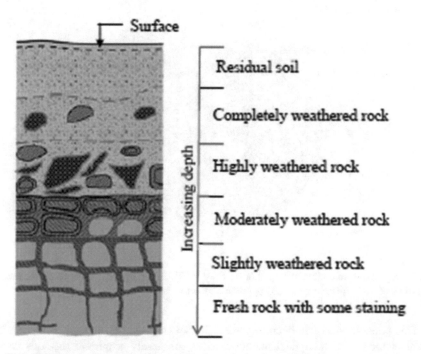

Fig. 4. Typical rock weathering profile from the surface (after Rhardjo et al. 2004).

2.2 Stress Induced Instability

During the course of excavation in the rock mass the in situ stresses are redistributed at the periphery of the tunnel boundary (Hoek and Brown 1980). When the in situ stresses are more than rock mass strength overstress lead to instability of underground opening occur. Mainly two types of instability occur due to induced stresses.

Rock Burst/Rock Spalling
Any sudden explosion due to release of strain energy in the strong and brittle rock masses is a phenomenon of rock burst. As per Barton et al. (1974), rock burst can be predicted σ_θ/q_c where q_c is the uniaxial compressive strength of rock and σ_θ tangential stress of rock mass. When σ_θ/q_c is more than 1.0 treated as heavy rock burst.

Tunnel Squeezing
The squeezing is nothing but mobilised elasto-pastic pressure due to failure of weak and fragile rock such as phyllite, shale and slate under the influence of tectonic stresses and over burden above the tunnel. This occur when the tangential stress at the perpheri of the tunnel opening exceeds the unconfined compressive strength of rock mass. Due to gradual formation of cracks along the foliation plane a zone of fractured rock mass is formed in that case the induced tangential stresses are travel out of the plastic zone Bray (1967) as shown in Fig. 5.

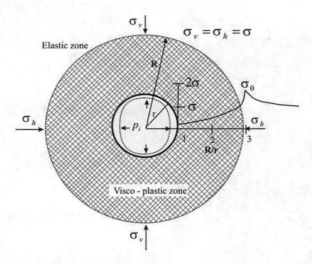

Fig. 5. Squeezing in a circular tunnel (after Bray 1967). In the figure, r is the tunnel radius, R is the radius of visco-plastic zone and p_i is the support pressure.

This phenomenon has been explained by several authors by empirical, semianalytical methods and analytical methods. Most commonly empirical methods such as Jethwa et al. (1981), Dube et al. (1986), Singh et al. (1992), Aydan et al. (1993), Goel et al. (1995), Grimstad and Barton (1993) during the past years. Singh et al. (1992) differentiated squeezing and non-squeezing using overburden height and Q-rating. Similarly several authors defined squeezing based on other criteria as well. Adyan et al. (1993), Hoek and Marinos (2000), Kovari (1998) explained the squeezing condition by semi analytical methods and analytically such as Carranza-Tores and Fairhust (2000).

Hoek and Marinos (2000) defined squeezing based on strain % (which is tunnel closure/tunnel diameter × 100) and ratio of rock mass strength and in situ stress. A classification was proposed for squeezing levels in which strain varying from 1% to 10% as shown in Fig. 6. Strain 10% indicates high squeezing conditions whereas at 1% strain the squeezing is minor. And if the strain % is <1% it indicates few support problems. In this study this classification has been used to define the squeezing problems.

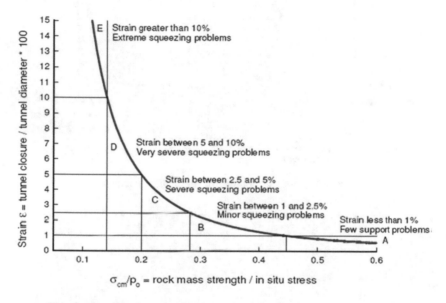

Fig. 6. Classification of squeezing behaviour (Hoek and Marinos 2000).

3   The Rohtang Tunnel

The proposed 8.8 km long two lane bi directional Rohtang highway tunnel above 3000 m from MSL across Rohtang Pass near Manali in Kullu district is of strategic importance from defence point of view as it will provide all weather connectivity between Manali and Keylong, the head quarter of Lahul-Spiti district.

The proposed 8.8 km long two-lane bidirectional Rohtang highway tunnel constructed to bypass Rohtang Pass in the eastern Pir Panjal range of the Himalayas on the Leh-Manali Highway above 3000 m from MSL as shown in Fig. 7. It is almost straight and runs along almost parallel to NorthEast-SouthWest direction. The tunnel overburden at South portals is about 50 m and about 100 m at North Portal. About 4 km of the tunnel length has an overburden of less than 600 m, about 900 m of length have overburden between 600–900 m, whereas about 3 km of tunnel have overburden between 900–1600 m and the remaining 750 m of tunnel have overburden more than 1600 m. The maximum overburden is approximately 1900 m. The excavation method is NATM with drill-blast. The layout and geological section of tunnel is shown in Figs. 8 and 9.

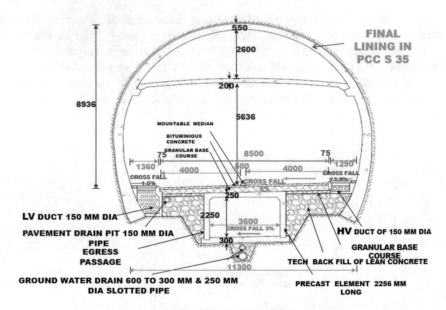

Fig. 7. Cross section of Rohtang highway tunnel

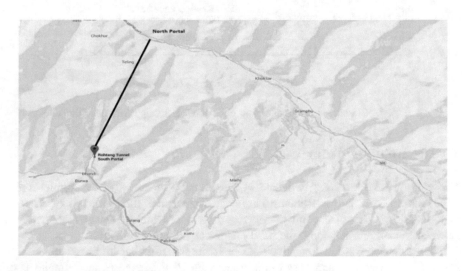

Fig. 8. Layout plan of Rohtang tunnel

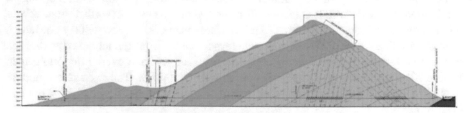

Fig. 9. Geological section of Rohtang tunnel

3.1   Geological Setup

The Rohtang tunnel is situated in Higher Himalaya and about 100 km south of Indus Suture Zone (Indus Tsangpo Ophiolite belt). The project is situated in Rohtang Axial zone which has a NW-SE trend which divides the drainage system of Chandra in the north and river Beas in the south with following tectonic features. The major tectonic feature separate the Lahaul-Spiti segment of the Spiti-Zanskar basin from the Tandi basin and also the Chamba-Bhadarwah basin and represents crystalline core with doubly plunging disposition. This zone has the elements of mantled gneiss exhibiting anataxis and migmatization and underlies the Infra cambrian sediments. The overlying sedimentary sequence and its metamor-phosed basal part are welded to the crystalline basement and it forms the root zone area of the Salkhala Nappe.

The area is always covered by snow on the top with valleys and valley slopes having thick cover of vegetation. Seri nala is a significant stream which crosses the alignment and joins Beas Kund nala at about 1.8 km downstream of the south portal site. The general drainage pattern of the area encompassing the project site is sub-parallel and is structurally controlled. The South and North portal of tunnels are at elevation 3060 M and 3071 M respectively. The general strike of foliation of these folded and faulted mountains is NW-SE. In general the average trend of dip direction and amount of foliation of these rocks is $220/35^0$ along with three major joints and one random joint. The main lithological rock units are Phyllitic quartzite, Quartzitic Phyllite, Migmatitic gneiss, Phyllites and Biotite Mica Schist. Anticipated rock type at different chainage of the tunnel with maximum and minimum overburden is presented in Table 3 and the anticipated squeezing sections in tunnel are presented in Table 4.

Table 3. Anticipated rock types

From	To	Length (m)	Min and max overburden (m)	Anticipated rock type
0	2000	2000	48.73 370.90	Schist/Phyllite
2000	2900	900	232.14–550.63	Migmatite/Fault zone
2900	4200	1300	550.63–1119.81	Schist/Phyllite
4200	8800	4600	102.82–1859.79	Migmatite with minor schist

Table 4. Predicted squeezing sections

From	To	Length (m)	Min and max overburden (m)	Rock type encountered
2100	2600	500	230–336	Schist/Phyllite
3700	4200	500	1089–1120	Schist/Phyllite
4200	8200	4000	501–1860	Migmatite with minor schist

3.2 Geotechnical Parameters

The geotechnical investigation were carried out through surface mapping, drifting and drilling of bore holes but being limited due to high overburden depth and snow covered area. The tunnel has been classified into classes on the basis of these investigations. To determine the physical and mechanical properties of these rocks laboratory tests were conducted, on rock samples collected from site during construction, by following Indian and International standards and practices. Rock mass properties and joint parameters were also determined through appropriate laboratory and field tests. Geotechnical parameters adopted for analysis of ground response are presented in Table 5. The rock mass properties for different rock type and overburden are determined by using RocLab software of Rocscince.

4 Analytical Approaches for Ground Response Analysis

Limited equilibrium method and Numerical Methods are the main analytical approaches used for the prediction and assessment of stresses and displacements in a circular tunnel excavated at great overburden depth.

The basic assumptions in closed form solutions are that the tunnel to be circular and to consider the rock mass subjected to a hydrostatic in situ state of stress, in which the horizontal and the vertical stresses are equal.

The lateral coefficient K is 1.5 where overburden are less than 900 m and 1.0 where overburden more than 900 m. As in a fault or shear zone, since rock has already undergone failure and is incapable of sustaining significant stress differences. Hence, even if the far field stresses are asymmetrical, the stresses within the fault zone are likely to be approximately hydrostatic (Hoek 1999) so laterial coefficient K is taken equal to 1.0.

Depending upon geological conditions numerical methods utilize continuous and/or discontinuous approach for rock mass modelling. In Continuum approach, the rock mass is considered as a continuum and theory of elasticity/ plasticity is applied to model the material behaviour. The commonly used continuum methods are Finite difference method (FDM), finite element method (FEM) and boundary element method (BEM). Whereas the discontinuums approach consider the rock mass as an assemblage of independent blocks or particles (Goodman and John 1977). The most commonly used method for this category is the discrete element method.

In this study the ground response was obtained using FLAC method. The Fast Langrangian Analysis of Continua (FLAC) is a two dimensional explicit finite difference program. The model was setup by specifying the material properties and initial boundary conditions. The procedure was adopted as given in FLAC manual version 7.0 and Mohr-Coulomb failure criteria was adopted for calculations.

The discretized model of the Rohtang tunnel developed by FLAC is shown in Figs. 10 and 11 for both shape of tunnel. The excavation diameter of Rohtang tunnel is 13 m. Considering the diameter the model geometry is created. The model dimensions are 150×150 m^2 which is more than 5 times the diameter of tunnel. The boundary of the model is chosen at a sufficient distance away from the excavation area to eliminate

Table 5. Geotechnical design parameters

Summary of results Mica Schist, Migmatic Gneiss, Phyllitic Quartzite and Quartzitic Phyllite

Name of test	Mica Schist	Migmatic Gneiss	Phyllitic Quartzite	Quartzitic Phyllite
Bulk density (gm/cc) dry	2.72	2.69	2.70	2.73
Bulk density (gm/cc) Saturated	2.74	2.69	2.70	2.74
Water content (%)	0.64	0.23	0.18	0.18
Porosity (%)	1.71	0.61	0.48	0.49
Specific gravity	2.82	2.73	2.77	2.73
Resistivity (Ω - m) dry	1.16×10^6	2.97×10^6	2.77×10^7	6.85×10^8
Resistivity (Ω - m) saturated	3.42×10^3	5.39×10^3	6.51×10^3	7.80×10^3
Wave velocity (km/s) dry	2.92	3.28	4.23	4.55
Wave velocity (km/s) saturated	2.02	2.52	3.65	3.99
Modulus (GPa) dry	23.19	28.94	48.31	56.52
Modulus (GPa) saturated	11.18	17.08	48.31	43.62
Index quality (%) Dry	48.67	54.67	70.50	75.83
Index quality (%) saturated	33.67	42.00	60.83	66.50
Slake durability (%)	98.06	98.04	98.93	98.78
Permeability (m/s)	9.00×10^{-7}	8.99×10^{-7}	2.77×10^{-8}	2.74×10^{-8}
Tensile strength (MPa) dry	8.19	6.06	12.28	11.71
Tensile strength (MPa) saturated	6.85	5.14	11.20	9.56
Unconfined compressive strength (MPa) dry	67.64	46.36	105.96	112.18
Unconfined compressive strength (MPa) saturated	60.43	44.22	94.49	83.26
Modulus of elasticity (GPa) dry	49.32	23.19	42.91	43.57
Modulus of elasticity (GPa) Saturated	48.05	15.05	41.10	43.28
Poisson's ratio (%) dry	0.25	0.26	0.26	0.24
Poisson's ratio (%) saturated	0.21	0.24	0.23	0.30
Material constant Hoek-Brown (mi)	10.00	30.04	19.12	20.60
Cohesion c (MPa)	5.31	9.20	18.15	16.34
Angle of internal friction ɸ in degree	58.45	50.44	48.68	49.00

the boundary effect. The boundary condition are applied in terms of both stresses and displacement. All the boundaries are fixed in X and Y directions due to overburden more than the boundary dimension.

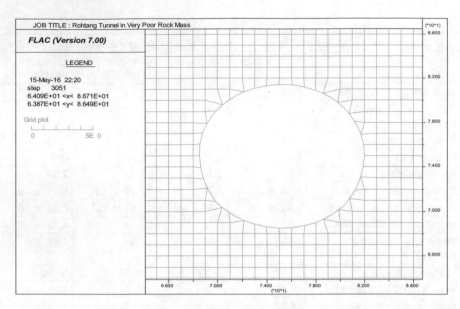

Fig. 10. Discretize model of circular section of tunnel

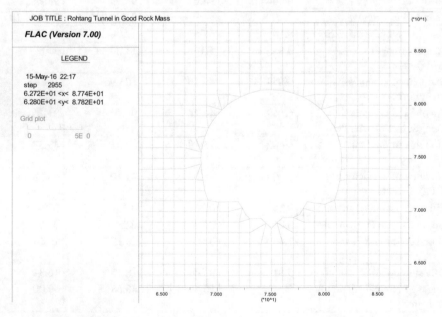

Fig. 11. Discretize model of horse shoe section of tunnel

5   Results and Discussion

Analysis based on closed form solutions for the chainages of rock masses available along the Rohtang tunnel has been carried out and stresses and displacement are obtained. Variation of radial and tangential stresses with distance from the centre of the tunnel and also variation of displacements (radial) with distance are plotted for all rocks and different overburdens. Typical such variations for Phyllitic Quartzite are shown in Fig. 12 As per the figure, it is clear that the $\sigma_\theta = 10.04$ MPa occurs at 12.16 m from the centre of tunnel, whereas radial stress is zero at the boundary. Maximum deformation of 121.56 mm is observed at the tunnel boundary and also radius of plastic zone is 12.16 m. Results for the rocks at different sections are summarized in Table 6.

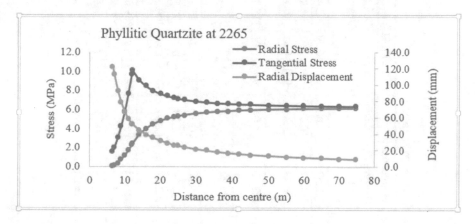

Fig. 12. Stress and deformation in phyllitic quartzite

On the basis of deformation, strain is calculated and squeezing condition is defined and is compared by empirical method given by Singh et al. and Goel et al. as shown in Table 3. Tunnel in the weak rocks such as Mica Schist is experiencing the excessive squeezing problem as compare to the Quartzitic Phyllite and Phylletic Quartzite. It can be observed in Table 6 the methods given by Singh et al. and Goel et al. are highly conservative and sometime gives squeezing for no squeezing condition. This happen because they work on empirical relations which have the statistical limitation.

Stresses and displacements are also obtained from FLAC and corresponding strains were obtained for all rocks at different overburden. The results are shown in Fig. 13(a), (b), (c) and (d) for Migmatic Gneiss and similar graphs were plotted for other rocks at different overburden as well. The results are compared in Table 7 along with the results obtained by closed form solutions. It is clear that the deformation values obtained from FLAC are higher than the closed form solutions. It is observed the results from the analytical study are matching with FDM results.

Table 6. Comparison of analysis

Location (m)	Rock. type	Rock class as per Q system	Our Burden (m)	Closed form of solution				FEM (FLAG) Analysis				Difference in deformation (%)
				Stresses (MPa)		Deformation (mm)	Strain (%)	Stresses (MPa)		Deformation (mm)	Strain (%)	
				σ_{tang}	σ_{rad}			σ_{tang}	σ_{rad}			
899	Mica Schist	Fair (V)	262.0	19.13	0	3.59	0.06	26.21	0.29	2.41	0.04	$K_o = 1.5$
1935	Mica Schist	Very poor (VII)	285.0	12.37	0	54.77	0.83	11.76	0.02	52.91	0.80	3
2205	Quartzitic phyllite	Extremely poor (VIII)	238.0	10.57	0	88.18	1.33	9.76	0.01	86.18	1.30	2.32
2265	Phyllitic Quartzite	Extremely poor (VIII)	234.0	10.04	0	121.56	1.84	9.46	0.01	115.80	1.75	4.97
2383	Rivet Bed Material	Exceptionally poor(IX)	232.0	9.02	0	381.22	5.77	8.43	0.01	347.9	5.26	9.57
5600	Mica Schist	Fair (V)	902.0	40.41	0	13.87	0.22	38.28	0.16	13.15	0.21	5.48
4100	Mica Schist	Very poor (VII)	1106.0	46.27	0	76.58	1.16	44.46	0.24	75.5	1.14	4.07
4600	Migmatic Gneiss	Poor (VI)	1173.0	51.98	0	102.67	1.64	49.23	0.64	100.57	1.61	2.08
5100	Migmatic Gneiss	Very poor (VII)	1860.0	79.96	0	366.16	5.54	74.75	0.82	339.00	5.13	8.01
8611	Migmatic Gneiss	Good (IV)	366.0	18.41	0	8.12	0.13	28.82	0.61	7.14	0.11	$K_o = 1.5$

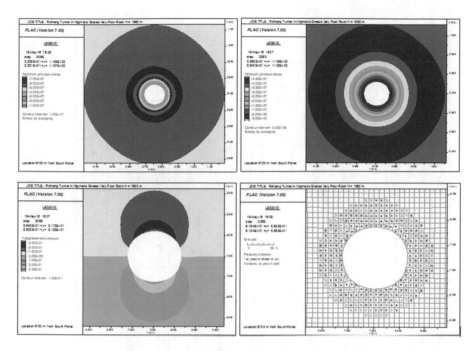

Fig. 13. (a) Maximum principal stress (b) Minimum principal stress (c) Y displacement (d) Plasticity indicator contours for Migmatic Gneiss at 1860 m of over burd

Stress concentration and excessive deformation is observed at sharp edges of the tunnel. One should always avoid sharp corners while designing the cross-section of the tunnel. When the water head was taken in the analysis, it was observed that the deformation on the opening was excessive and the FDM model was not even converging because of that excessive deformations. True Deformation due to water head was not calculated very precisely because the convergence issue.

Tunnel in the weak rocks such as Mica Schist is experiencing the excessive squeezing problem as compare to the Quartzitic Phyllite and Phyllitic Quartzite. Later rocks are having more UCS as compare to the former one.

Rock mass response behaviour from closed form solutions and FLAC obtained for all rock masses. FLAC software is used to stabilise the tunnel at all sections with shotcrete lining steel sets and rock bolts of appropriate input parameters. After installation of support shortcrete lining and steel sets are checked in bending and direct stresses and rock bolts are checked in tensile stresses.

Bending movements, axial force and structural displacements in supports for all rock types at different overburden are plotted and typical results obtained for shale are presented in Fig. 14(a), (b) and (c). Comparison of results without and with support is given in Table 8 for all sections.

Table 7. Details of stresses

Location (m)	Water condition	Shotcrete lining				Steel set embedded in concrete				Rockbolts		
		Direct stress (MPa)	Bend ing stress (MPa)	Safety	Structural displacement (mm)	Direct stress (MPa)	Bend ing stress (MPa)	Safety	Structural displacement (mm)	Tensile force (KN)	Safety	Structural displacement (mm)
899	Without water	8.18	0.26	Safe	1.23	–	–	–	–	–	–	–
1935	Without water	21.84	3.07	Safe	10.35	21.23	9.86	Safe	5.89	186.9	Safe	21.49
	With water	22.87	5.07	Safe	13.02	23.2	12.88	Safe	7.56	238.8	Safe	24.28
2205	Without water	22.7	3.07	Safe	12.93	21.24	16.21	Safe	7.05	231.4	Safe	32.48
	With water	22.72	3.00	Safe	12.70	22.02	13.48	Safe	7.00	243.9	Safe	32.36
2205	Without water	21.6	3.29	Safe	9.96	17.28	9.19	Safe	4.70	217.0	Safe	31.02
	With water	21.41	3.64	Safe	14.31	23.19	14.19	Safe	8.33	266.0	Safe	34.58
2265	Without water	23.22	0.23	Safe	17.75	17.73	20.22	Safe	9.49	151.8	Safe	25.99
	With water	23.6	0.5	Not Safe	14.52	14.06	15.96	Safe	7.49	170.8	Safe	25.14
3600	Without water	9.94	1.65	Safe	2.97	–	–	–	–	128.3	Safe	9.59

(continued)

Table 7. (*continued*)

Location (m)	Water condition	Shotcrete lining				Steel set embedded in concrete				Rockbolts		
		Direct stress (MPa)	Bend ing stress (MPa)	Safety	Structural displacement (mm)	Direct stress (MPa)	Bend ing stress (MPa)	Safety	Structural displacement (mm)	Tensile force (KN)	Safety	Structural displacement (mm)
4100	Without water	18.27	2.97	Safe	5.80	–	–	–	–	267.3	Safe	34.37
4600	Without water	8.09	1.60	Safe	5.37	–	–	–	–	283.1	Safe	33.33
6100	Without water	4.86	0.88	Safe	1.45	–	–	–	–	465	**Not Safe**	106.6
8611	Without water	12.4	1.52	Safe	2.19	–	–	–	–	–	–	–

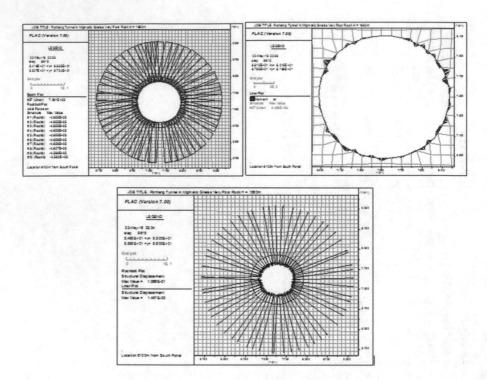

Fig. 14. (a) Axial force (b) Bending moment and (c) Displacement in support

Numerical analysis of support system at location 2383 chainage, which is in exceptionally poor rock. The tunnel is supported by 450 mm shotcrete, ISMB350 embedded in concrete and 10 m long 25 mm diameter rock bolt with spacing 1 m × 1 m in dry condition was found to be safe, but when pore pressure of 232 m head is applied the support system fails even after applying secondary support system, which is having 500 mm thick concrete lining. So, it concludes that proper drainage of pore pressure is needed to support such grounds. At chainage 6100 where the overburden is 1860 m is unsafe even after applying 12 m long 32 mm diameter with 0.75 m spacing having rock bolt displacement 106 mm, in such type of high squeezing ground condition various authors recommends sufficient ductile support to accommodate high volume change such as light wire mesh with lattice girder. For shear zone condition, forepoling and advance face stabilization are required.

Table 8. Comparison of deformations

Location (m)	Rock type	Rock class as per Q system	Over burden (m)	Without support			After support		
				R_o(m)	Deformation (mm)	Strain	R_o(m)	Deformation (mm)	Strain (%)
899	Mica Schist	V	262.0	7.5	2.41	0.04	7.5	2.14	0.15
1935	Mica Schist	VII	285.0	11.48	52.91	0.80	7.5	20.52	0.34
2205	Quartzitic Phyllite	VIII	238.0	10.46	86.18	1.30	6.68	34.40	0.52
2265	Phyllitic Quartzite	VIII	234.0	11 45	115.8	1.75	6.68	31.6	0.50
2383	River Bed Material	IX	232.0	24.45	347.9	5.26	7.47	26.38	0.40
3600	Mica Schist	V	902.0	9.49	13.15	0.21	9.49	11.95	0.40
4100	Mica Schist	VII	1106.0	12.47	75.5	1.14	11.5	54.37	0.90
4600	Mgmatic Gneiss	VI	1173.0	9.47	100.57	1.61	8.43	84.4	3.17
6100	Mgmatic Gneiss	VII	1860.0	11 36	339	5.13	10.37	270.0	4.29
8611	Mgmatic Gneiss	IV	366.0	6.72	23.2	0.11	6.72	20.8	0.33

6 Conclusions

The Rohtang tunnel in the Himalayas traverses through much diversified geology experiencing high ground stresses. Time dependent behaviour of tunnels in squeezing rock is investigated using closed form solutions and FLAC 2D methods. Stresses and displacements are obtained before and after the installation of support measures. Only shotcrete and rock bolting proves ineffective to control squeezing in such type of rock, hence advance face stabilization suggested approaches like fore poling, lining stress controller, lattice girder in these sections are required.

References

Ajalloeian, R., Lashkaripour, G.: Strength anisotropies in mud rocks. Bull. Eng. Geol. Environ. **59**, 195–199 (2000)

Aydan, O., Akagi, T., Kawamoto, T.: The squeezing potential of rocks around tunnels, theory and prediction. Rock Mech. Rock Eng. **26**(2), 137–163 (1993)

Barton, N., Lien, R., Lunde, J.: Engineering classification of rock masses for the design of tunnel support. Rock Mech. **6**, 189–236. Springer-Verlag (1974)

Barton, N., Bandis, S., Bakhtar, K.: Strength, deformation and conductivity coupling of rock joints. Int. J. Rock Mech. Min. Sci. Geomech. Abstr. **22**, 121–140 (1985)

Barton, N.: Some new Q-value correlation to assist in site characterization and tunnel design. Int. J. Rock Mech. Min. Sci. **39**, 185–216 (2002)

Beavis, F.C., Roberts, F.I., Minskaya, L.: Engineering aspects of weathering of low grade metapelites in an arid climate zone. Quat. J. Eng. Geol. Lond. **15**, 29–45 (1982)

Bieniawaski, Z.T.: Determining rock mass deformability: experience from case histories. Int. J. Rock Mech. Min. Sci. Geomech. Abstr. **15**, 237–247 (1978)

Bieniawskiien, Z.T.: Engineering Rock Mass Classifications. Wiley, New York (1989)

Bieniawaski, Z.T.: Classification of rock masses for engineering: the RMR-system and future trends. In: Hudson, J.A. (ed.) Comprehensive Rock Engineering, vol. 3, pp. 553–573 (1993)

Bray, J.W.: A study of jointed and fractured rock. Theory of limit equilibrium. Felsmechanik und ingeneurgeologie (Rock mechanics and engineering geology), vol. 5, pp. 197–216 (1967)

Carranza-Torres, C., Fairhurst, C.: Application of the convergence confinement method of tunnel design to rock masses that satisfy the Hoek-Brown failure criteria. Tunn. Undergr. Space Technol. **15**, 187–213 (2000)

Deere, D.U., Hendron, A.J., Patton, F.D., Cording, E.J.: Design of surface and near surface construction in rock. In: 8th US Symposium on Rock Mechanics, Failure and Breakage of Rock, pp. 237–302 (1967)

Dershowitz, W.S., Baecher, G.B., Einstein, H.H.: Prediction of rock mass deformability. In: Proceedings of 4th International Congress on Rock Mechanics, Montreal, Canada, vol. 1, pp. 605–611 (1979)

Dube, A.K., Singh, B., Singh, B.: Study of squeezing pressure phenomenon in tunnel Part-I and Part-II. Tunn. Ling Undergr. Space Technol. **1**(1), 35–39 (Part-I) and 41–48 (Part-II), USA (1986)

Goel, R.K., Jethwa, U.L., Paithakan, A.G.: Tunnelling through young Himalayas - a case history of the Maneri - Uttarkashi power tunnel. Eng. Geol. **39**, 31–44 (1995)

Grimstad, E., Barton, N.: Updating the Q-system for NMT. In: Proceeding of the International Symposium on Sprayed Concrete – Modern Use of Wet Mix Sprayed Concrete for Underground Support, Fagernes. Norwegian Concrete Association, Oslo, Norway (1993)

Gupta, A.S., Rao, K.S.: Weathering effects on the strength and deformational behaviour of crystalline rocks under uniaxial compression state. Eng. Geol. **56**, 257274 (2000)

Hawkins, A.B.: Aspects of rock strength. Bull. Eng. Geol. Environ. **57**, 17–30 (1998)

Hoek, E., Brown, E.T.: Underground Excavation in Rock, p. 527. Institute of Mining and Metallurgy, London (1980)

Hoek, E.: Strength of rock and rock masses. ISRM News J. **2**, 4–16 (1994)

Hoek, E., Brown, E.T.: Practical estimates of rock mass strength. Int. J. Rock Mech. Min. Sci. **34**, 1165–1186 (1997)

Hoek, E., Marinos, P.: Predicting tunnel squeezing problems in weak heterogeneous rock masses. Tunn. Tunn. Int. **32**(11 and 12), 45–51 and 34–36 (2000)

Hoek, E., Torres, C.C., Corkum, B.: Hoek-Brown failure criterion – 2002 edition. In: Proceedings North American Rock Mechanics Society Meeting. Toronto, Canada (2002)

Hudson, J.A.: Rock Mechanics Principles in Engineering Practice. CIRIA Ground Engineering Report: Underground Construction. Butterworths, London, 72 p (1989)

Hudson, J.A., Harrison, J.P.: Engineering Rock Mechanics. An Introduction to the Principles. Pergamon, Oxford, 444 p (1997)

ISRM: Report of the commission on terminology. Lisbon (1975)

ISRM: Suggested methods for the quantitative description of discontinuities in rock mass. Int. J. Rock Mech. Min. Sci. Geomech. Abstr. **15**, 319–368 (1978)

ITASCA Inc. FLAC [2D] Fast Lagrangian Analysis of Continua in 2 dimensions. Version 7.00, FLAC [2D] manual 2011

Jethwa, J.L.: Evaluation of Rock Pressures in Tunnels through Squeezing Ground in Lower Himalayas. Ph.D thesis, Department of Civil Engineering, University of Roorkee, India, 272 (1981)

Kovari, K.: Tunnelling in squeezing rock. Tunnel **5**, 12–31 (1998)

Nasseri, M.H.B., Rao, K.S., Ramamurthy, T.: Anisotropic strength and deformational behavior of Himalayan schist. Int. J. Rock Mech. Min. Sci. **40**, 3–23 (2003)

Nilsen, B., Palmstrøm, A.: Hand Book No. 2. Engineering geology and rock engineering. Norwegian group for rock mechanics in cooperation (NBG) with Norwegian tunnelling society (NFF), Norway, 249 p (2000)

Palmstrøm, A.: RMi – a rock mass characterization system for rock engineering purposes. Ph.D thesis. Department of geology, University of Oslo, 400 p (1995)

Ramamurthy, T.: Strength and modulus responses of anisotropic rocks. In: Hudson, J.A. (ed.) Comprehensive Rock Engineering, vol. 1, pp. 313–329. Pergamon (1993)

Rhardjo, H., Aung, K.K., Leong, E.C., Rezaur, R.B.: Characteristics of residual soils in Singapore as formed by weathering. Eng. Geol. **73**, 157–169 (2004)

Rocscience: RocLab software program – free download from the Rocscience website (2002). www.rocscience.com

Serafim, J.L., Pereira, J.P.: Consideration of the geomechanics classification of Bieniawaski. In: Proceedings: International Symposium on Engineering Geology and Underground Construction, pp. 1133–1144 (1983)

Singh, B., Jethwa, J.L., Dube, A.K., Singh, B.: Correlation between observed support pressure and rock mass quality. Tunn. Undergr. Space Technol. **7**, 59–74 (1992)

Engineering Failure Analysis and Design of Support System for Ancient Egyptian Monuments in Valley of the Kings, Luxor, Egypt

Sayed Hemeda[1,2(✉)] [iD]

[1] Conservation Department, Faculty of Archaeology, Cairo University,
Giza, Egypt
sayed.hemeda@cu.edu.eg
[2] Civil Engineering, Aristotle University of Thessaloniki,
12613 Thessaloniki, Greece

Abstract. The main objectives of the experimental and numerical geo-environmental and geotechnical simulations and analyses carried out in the present study are to investigate the static stability, safety margins and engineering failure of the tomb of Sons of Ramsses II (KV5), under their present conditions, against unfavorable environmental circumstances (i.e. extensive weathering due to water and flash floods impact in the past and present), utter lack of preservation, geostatic overloading of structural rock support pillars, geotechnical and extreme seismic conditions. Also to design an appropriate geotechnical support system, according to the engineering rock mass classification, in particularly the rock mass rating RMR and quality rock tunneling index Q-system. Based on the underground engineering stable equilibrium theory and rock mass classification, three support structure techniques are provided and detailed illustrated with the case of KV5 in this study.

Keywords: Geotechnical problems · Rock character · Support structure
Tomb of the Sons of Ramses II · Valley of the Kings

1 Introduction

The paper represents the first comprehensive experimental and numerical study for engineering failure analysis and appropriate design for the permanent mechanical support system for the tomb of the Sons of Ramesses II (KV5).

During the late 18th Dynasty and throughout the19th, the tombs are usually located further down the Valley some distance from the rock walls. The builders often quarried through talus slopes, such as in the case of the tomb of Sons of Ramses II. It is clear that the tomb of sons of Ramsses II is much more susceptible to surcharge geostatic loading from the overburden rock strata, rock bursting, and structural damage of support pillars and walls induced by the water and past/recent flash floods impacts caused by heavy rain in the Valley. Since some of this tomb also makes contact with the underlying shale layers, which have the potential for swelling and shrinkage under

© Springer Nature Switzerland AG 2019
M. Badr and A. Lotfy (Eds.): GeoMEast 2018, SUCI, pp. 126–158, 2019.
https://doi.org/10.1007/978-3-030-01884-9_10

changing moisture conditions. Expansive damages to these underground structures have been widely noticed in the Valley of the Kings. This tomb has a tendency to be the most noticeably awful safeguarded tomb in the Valley of the Kings. The Esna shale in the valley is especially powerless and flimsy. It postured issues to the antiquated quarryman, as well as to the cutting edge conservator also. At the point the shale expands when it is exposed to moisture increase, and its expansion can tear a hill side apart.

The engineering analysis had been carried out through the following four steps: 1-Evaluation of the surrounding rocks (marl limestone) by experimental investigation and the RocLab program to obtain Hoek Brown Classification criterion, Mohr–Coulomb fit and the rock mass parameters in particular the global strength and deformation modulus. 2-Qualitative and quantitative estimations of relevant factors affecting the stability of the tomb in particularly the overburden or geostatic and dynamic loading. 3-2D and 3D integrated geotechnical modeling of the tomb environment for stress, displacement analyses and determination of volumetric strains and plastic points using advanced codes and programs like Examine 2D and PLAXIS 3D. The numerical analysis results indicated that the safety factor of the rock pillar structural supports is 1.37 and the overstress state is 1.28 MPa. 4-Remedial and retrofitting policies and techniques, static monitoring and control systems which are necessary for the strengthening and stability enhancement of the tomb, where the rock mass classification indicated that the rock mass where the KV5 is excavated is poor rock, with RMR 39 and Q value 1.87.

Amongst the many monument types which exist all over the world, underground sites such as caves, tombs, crypts and catacombs can be singled out as a category which has its own particular set of "adversaries". These locations are to a certain extent "protected" by the earth or rock surrounding them; this is especially so when these sites remain sealed, or has only one small or partially blocked opening to the exterior. However, when an interred site is discovered and uncovered, its microclimate is disturbed and fluctuations in internal conditions commence. These variations become accentuated if the protective covering is removed either during excavation or later to create a new wider access to the site. This instability eventually leads to deterioration of the site and in particularly any decorations or paintings it may contain. Further deterioration is caused by other unrelated sources such as water seepage and occasionally also flooding like the tomb KV5 which under investigation, salt damage and the accumulation of dust, debris and other contaminants. The problems are common to all painted underground and semi-buried sites in the valley of kings at Luxor, Egypt.

It is important to say that in geological engineering, including in underground rock engineering and rock mechanics, lots of the hazards sources arise from geotechnical uncertainty or error. The sources of uncertainty can be classified as: (1) inalienable spatial and fleeting fluctuation; (2) estimation and observing blunders; (3) demonstrating vulnerability; (4) load and stresses vulnerability (Brown 2012). In geotechnical engineering it is perceived that stone disfigurement is imperative in deciding the advancement of characteristic structures and structural highlights. Numerous investigations and field work has been done to comprehend the fragile break procedures and systems. Quite a bit of this focused on the research center testing and the estimation of fragile crack limits (Deere and Varde 1990).

The Geotechnical instability problems and degradation phenomena of rock cut tombs in the Valley of the Kings (KV) is likely to be dominated by gravity fall and sliding on structural features, also other factors such as excessively high rock stress, creep effect, poor geotechnical properties of rock structures, weathering and/or swelling rock and flash floods caused by heavy rains in the Valley, vibrations and dynamic loading as well as utter lack of preservation become important and can be evaluated by means of a classification of rock quality. The Esna shale in the valley is particularly weak and unstable. It not only posed problems to the ancient quarryman, but to the modern conservator as well. When the shale comes into contact with moisture, it expands and can literally tear a hill side apart.

The tomb was looted in ancient times. From that point forward, it has been hit by no less than eleven glimmer surges caused by overwhelming downpours in the Valley. These have totally filled the tomb with trash and truly harmed its completely embellished dividers. From around 1960–1990, visit transports stopped over the tomb; their vibrations made genuine harm parts of the tomb close to the roadway, as completed a spilling sewer line introduced over the passage when the Valley of the Kings rest house was assembled.

In October and November of 1994, two flash floods happened in the Valley of Kings, sending a notice to all legacy directors. In the two cases, a nearby desert rainstorm happened in the region of the Valley of Kings. Tempest water spillover and residue entered the tomb of Sons of Ramsses II and other numerous o tombs and caused disintegration of crevasse floors.

Present day farming procedures have likewise added to the land properties of the Nile Valley bowl. Today, ground water levels have ascended around there, and debilitate low lying shaft tombs and the funeral home sanctuaries on the edge of the development, and in addition the outstanding Luxor and Karnak sanctuaries on the east bank. The tomb of Sons of Ramsses II (KV5) is located at the middle of the Valley of the Kings, East Valley, Thebes West Bank at Thebes. The Theban Mapping Project's excavations have shown that KV 5 contains not just the six rooms first seen by Burton in 1825, but over 150 corridors and chambers dug deep into the hillside, as shown in Fig. 1.

KV 5 itself is the largest rock cut tomb in the Valley of the Kings; pillared chamber 3 is the largest chamber of any tomb in the Valley of the Kings. Chambers 1–6 had been discovered in 1825 by James Burton, all other had been discovered by Theban Mapping Project in 1995 (Weeks 1992, 1994 and Weeks 1995), as shown in Fig. 2.

There is an adjustment in the tomb's essential pivot after chamber 3; a few chambers lie underneath different chambers; two hallways reach out toward the northwest underneath the passageway and the street before the tomb. Pillared chamber 3 has more columns (sixteen) than some other chamber in the Valley of the Kings. The measured dimensions of the KV5 are maximum height of 2.85 m, width of 0.61 ~ 15.43 m, total length of 443.2 m; total area of 1266.47 m^2 and total volume of 2154.82 m^3. Pillars Conditions are excavated, decoration damaged, damaged structurally (Weeks 1998, 2000, 2006), as shown in Fig. 3.

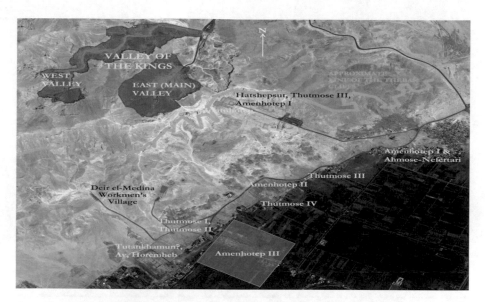

Fig. 1. Aerial photograph indicate the east (main) Valley of the Kings (KV) at Luxor Egypt. Modified after Google earth map.

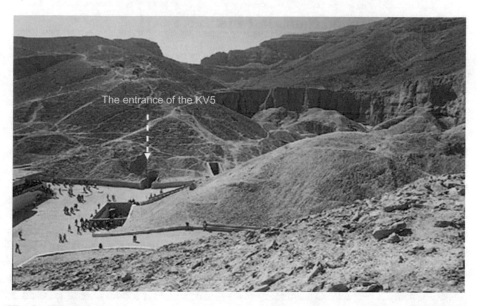

Fig. 2. Location of the tomb of Sons of Ramsses II (KV5) at the east (main) Valley of the Kings (KV), Luxor Egypt.

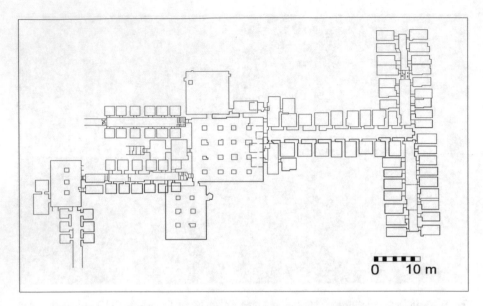

Fig. 3. The present layout and plan of the tomb of Sons of Ramsses II (KV5).

2 Experimental and Numerical Methods

The rock mass petrography and mechanical strength where the tomb of sons of Ramses II is excavated has been analyzed by experimental investigations, which include XRD, XRF analysis and thin section examination under polarized light microscope. A comprehensive program for petro physical and mechanical testing includes the uniaxial compression test and ultra-sonic wave velocity through the materials (PUNDT) has been established. The RocLab program has been utilized to calculate the Hoek–Brown Classification and criterion also to calculate the Mohr–Coulomb fits and rock strength parameters in particularly the deformation modulus. Underground structures security investigation is performed utilizing the limited component (FE) technique. The exploration shows a far reaching study for the stone cut tombs security investigation. The security examination incorporates a disappointment examination as well as the impact of weathering, specifically the materials wear on the differential settlement have been researched. Inspect 2D is a limited component program created for numerical investigation of geotechnical and underground and underground structures.

The deformation of these rock cut tombs has been computed as realistically as possible, using a progressed nonlinear elasto-plastic material model should be used in PLAXIS 3D which is equipped for using such propelled material models. 3D Plastic model is utilized for disfigurement and union investigation in this examination. The solidification examination is performed utilizing PLAXIS 3D. Likewise in this exploration, the author attempted to construct and investigate a three-dimensional (3D) limited component demonstrate (FEM)of the pillared chamber 3 with its structurally damaged sixteen rock pillars and the large northern hall which are excavated in this poor and extensively weathered marl limestone deposit (member 1), using the PLAXIS 3D code.

The Rock Mass Classification calculations are utilized for the general assessment of the rock mass where the KV5 is excavated. The results of the rock mas rating (RMR) and Q-system values were utilized to design an appropriate support system.

3   The Geology of Gebel El-Gurnah, Luxor

Gebel El-Gurnah is located some 4 km to the west of the River Nile, opposite to Luxor. The main exposed rock units in Gebel El-Gurnah are the Esna Shale and Thebes limestone formations. The tombs of the kings were excavated in the Thebes formations at northern side of Gebel El-Gurnah and the tombs of the queens were excavated at the southern side, (Dunn 2014; Wust and Macline 2000).

The main exposed rocks in Gebel El-Gurnah are the Esna Shale (late Paleocene-Early Eocene) and the conformatably overlying Thebes formation (Early Eocene).

Esna Shale: The lower 25 m of this formation is less calcareous, usually is green dark grey, and sometimes nearly block. The upper shale is whitish grey and greenish, more compact and carries more gypsum veinlets. Brownish red and yellow hematitic and limonitic concretions are present; the ferruginous concretions are characteristic feature foe the whole formation. The gypsum veinlets run mostly parallel to the bedding planes (Wust and Maclane 2000), as shown in Fig. 4a.

Fig. 4. (a) Esna Shale and (b) Marl Limestone (Member 1), Gebel El-Gurnah.

Thebes Formation: the Thebes formation exposed in the valley of kings could be subdivided into three members (from base to top) Hamadat, Beida and Al-Geer members however, the Thebes formation conformably overlying the Esna Shale. The lower member Hamadat is white, chalky indurated limestone with flint concretions, the middle member Beida is made up indurated, thick bedded, nodular limestone with flint bands extending parallel to the bedding planes, the uppermost member Al-Geer consists mainly of white limestone, (Aubry et al. 2008 and Silotti 1997), as shown in Figs. 4b and 5.

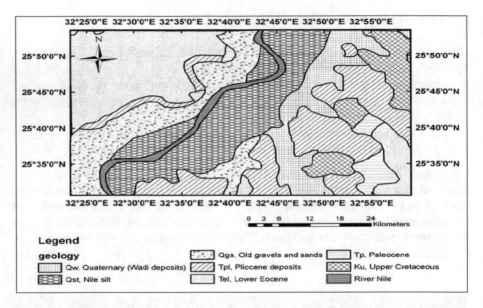

Fig. 5. Geological setting of the Valley of the Kings, Luxor, Egypt. (Geological Egyptian Authority).

There are many faults in the SW corner of the Valley of the Kings; it is very composite in its nature. Numbers of faults are cutting the Eocene limestone Formations. Typically, those issue dividers bring differentiated throughout sliding, and veins about crystalline calcite have developed in the interceding spaces. The calcite may be stringy Furthermore structures overstepping bundles, which provide for the course What's more, sense from claiming slip.

Ordinary faults, demonstrating level development for An NE-SW direction, are abundant. However, one substantial fault, on the Nw side of the valley, may be dominantly strike-slip (and left-lateral), while others need aid oblique-slip (left-normal, alternately right-normal). Despite those five faults that required been measured are not enough will a chance to be statistically significant, they are commonly perfect. As shown in Figs. 6 and 7. Figures 8 and 9 present the state of preservation of the KV5 and the geological and geotechnical induced rockmass stability problems. Where the brittle rock, high stress conditions lead to rockbursting (the sudden release of stored strain energy) bursts manifest themselves through sudden.

Fig. 6. Thebes Formations, Valley of the Kings, Luxor, Egypt.

Fig. 7. Rock structures such as joints, bedding's characters of the Valley of the Kings (KV).

Fig. 8. Extensive structural damage and engineering failure of the structural pillars, sidewalls and Ceiling of the Corridors and Chambers in the KV5.

Fig. 9. Brittle rock, high stress conditions lead to rockbursting (the sudden release of stored strain energy) bursts manifest themselves through sudden. (After TMP).

4   Results of Experimental Investigations

4.1   Geotechnical Properties of Intact Rock Specimens and Discontinuities

Twenty-three cylindrical rock specimens have been prepared from the surrounding rock and the supporting pillars to delineate the physical and mechanical properties. Specific gravity, unit weight, water absorption, porosity and degree of saturation are the physical aspects determined. While, the mechanical characterization included the determination of the uniaxial compressive strength, elastic static modulus of elasticity and Brazilian splitting tensile strength, as well as the Non-Destructive Ultrasonic Pulse Testing to the wave velocity through the brick specimens, the dynamic Young's modulus and shear modulus. All the soil/rock testing referring to the ASTM.

Thin-sections prepared on the limestone samples where the KV5 is excavated, refers that the limestone is fine-grained calcite, embedded in a micritic matrix rich in amorphous silica, fossils like Foraminifera and large grains of quartz.

The XRD analysis indicated that the major contents of Esna shale are quartz (SiO_2) and Montmorillonite ($Na_{0.2}$ $Ca_{0.1}$ Al_2 Si_4O_{10} $(OH)_2$. (H_2O) $_{10}$, the minor contents include the Kaolinite and Illite with Calcite traces. The bulk unit weight of the Esna shale is 1.79–1.86 g/cm^3, and the uniaxial compressive strength is 4.22–4.43 kg/ cm^2. As shown in Table 1.

Table 1. Geotechnical properties of the Esna shale samples (KV5).

Sample No	Natural water content %	γb N/m^3	γd N/m^3	qu MPa	OMC %
1	10	18	17	0.4	13
2	8.5	18.5	17.8	0.5	13.6
3	9	18.3	17.2	0.44	13.2

Petro-physical properties: Physical measurements referred that the unit weight (γ) of marl limestone of KV5 is between 20 and 21 kN/m^3, water absorptions (Wa) were between 10 and 12% and the apparent porosity (n) ranged from 14 to 19%.

Shear Wave Velocities (Vs): Shear wave velocities of limestone samples were measured by PUNDT (ASTM 597, ASTM D 2845-83). They varied from 0.7 to 1.0 km/s (with an average of 1 km/s for an orientation perpendicular on the bedding plane.

Uniaxial Compression Test: The compressive strength (σ_c) for the sidewalls is between 6 and 7 MPa, while the (σ_c) for the supporting rock pillars is 1 MPa because of the impact of the past and recent flash floods.

The static Young's modulus (E) = 10 GPa, Poisson Ratio (v) = 0.28–0.30, Fig. 10 shows the test set and the results are summarized in Tables 2 and 3.

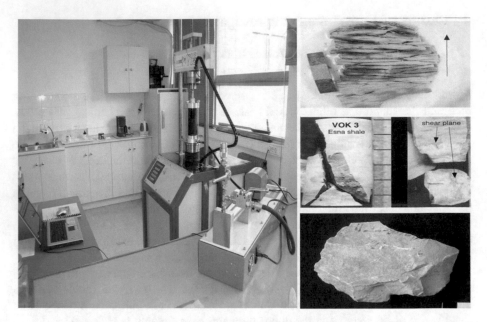

Fig. 10. Esna Shale and marl limestone samples under investigation.

Table 2. Geotechnical properties of the intact limestone samples (KV5).

No	PI (MPa	σc (MPa) Sidewalls	σc (MPa) Pillars	Vs (km/s)	RN
1	0.4	7.1	0.9	0.7	18
2	0.5	6.9	0.8	0.5	19
3	0.4	7.5	0.95	0.8	20
4	0.3	7.0	0.9	0.9	18
5	0.5	6.9	0.8	0.7	17
6	0.3	6.6	0.7	0.6	18
7	0.4	7.0	1.1	0.7	19

4.2   Analysis of Rock Mass Strength Using RocLab Program

RocLab is a software program for determining rock mass strength parameters, based on the latest version of the generalized Hoek–Brown failure criterion.

Hoek–Brown Classification: Intact uniaxial compressive strength of intact rock (σ_{ci}) = 7 MPa, GSI geological structure index = 50, intact modulus (mi) = 10, disturbance factor (D) = 0, intact rock deformation modulus Ei = 3500 MPa, modulus ratio (MR) = 500.

The generalized Hoek–Brown Criterion failure criterion: mb = 1.677, s = 0.0039. a = 0.506, where (s) and (a) are constants of the rock mass, calculated from the geological strength index (GSI) and disturbance factor (D).

Mohr–Coulomb Fit: Cohesion c = 0.349 MPa, Friction angle φ = 30°.

Table 3. a. Geotechnical properties of the intact limestone samples with depth and b. shear parameters of the discontinuities (KV5).

Depth	Weathering Grade	UCS (MPa)	E (MPa)
0–2 m	IV	1–5	2000
2–4 m	III	5–10	6000
4–6 m	II-III	10–11	10000
6–8 m	III	12–13	10000
Type	Peak friction	Residual friction	In-situ
Joints	30°	30°	JRC (L = 1 m) = 3–4
Joints	35°	25°	c = 30 kPa
			Φ = 35°
Joints	35°	30°	–

Rock mass parameters: Tensile strength of intact rock $\sigma_t = -0.016$ MPa, Uniaxial compressive strength, as shown in Figs. 11 and 12.

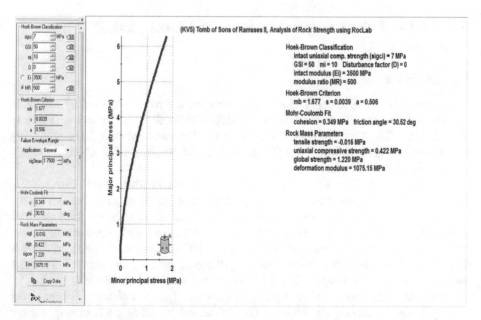

Fig. 11. Major and minor principal stress curve of marl limestone (KV5) using the RocLab program.

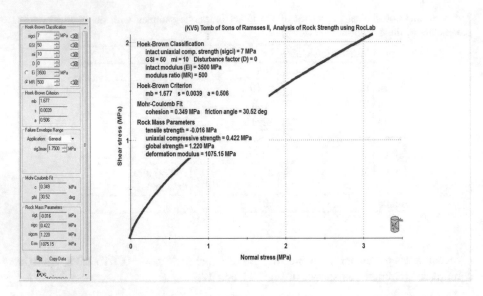

Fig. 12. Shear stress-Normal stress curve of Marl limestone (KV5), using the RocLab program.

5 Results of Thenumerical Analysis and Geotechnical Modeling

5.1 2D Static Analysis

In the initial 2D static analysis, the Sons of Ramses II tomb is modeled by assuming non-linear soil/ rock plastic model and the Mohr–Coulomb failure criterion, (Hemeda and Pitilakis, 2010), the 2D examine code is used for present study. The following parameters are used: $\varphi = 30°$, $c = 500$ kN/m^2, $E = 10.100E + 06$ KN/m^2, $v = 0.3$, $Vs = 800$ m/s for the rock material.

The results from the preliminary static analysis which are illustrated in Figs. 13, 14, 15, 16, 17, 18, 19 20 indicate that the maximum total displacements of the rock pillars in the large sixteen pillar chamber 3 were 1.2×10^{-4} m and the vertical displacements were small (of the order of millimeters 1.5×10^{-4} m), Horizontal displacement 1.25×10^{-5} m, the maximum volumetric strain is 3.5×10^{-5} m, and the spalling criterion is 0.22. While the maximum ground vertical displacements on the roof of the large western two halls were large 4.5×10^{-5} m and the volumetric strain is 7×10^{-6}.

The rock pillars in the sixteen pillars largest hall (pillared chamber 3) are under relatively high compression stresses. The calculated effective peak principal compressive stresses on supporting rock pillars are about 900 kPa. The maximum shear stress is 0.15 MPa, and the maximum shear strain is 1.3×10^{-5}.

For the large northern hall, The calculated effective peak principal compressive stresses is about 600 kPa but the maximum vertical displacement on the roof is too large 1.2×10^{-4} m and the maximum volumetric strain is 4.5×10^{-6}, the results of the mathematical modeling are represented in Figs. 13, 14, 15, 16, 17, 18, 19 20. Also the maximum vertical stress on the roofs and sidewall of Chamber 1 and Chamber 2

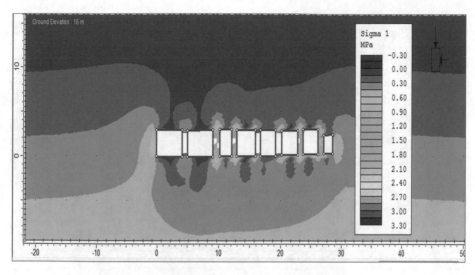

Fig. 13. Effective vertical stresses distribution through the rock pillars in Chamber 3. Examine 2D. The units of distance and depth in all figures are in meter.

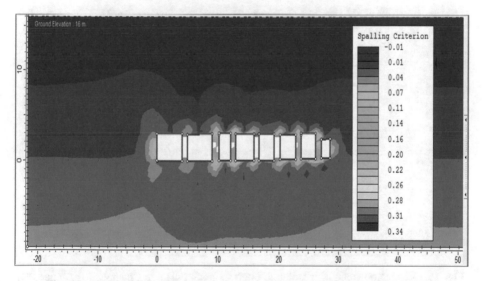

Fig. 14. Spalling Criterion through the rock pillars in Chamber 3. Examine 2D.

reached 350 kPa, and the maximum vertical displacement reached 4.5×10^{-5} m, as shown in Figs. 15 16, 17, 18, 19 20.

5.2 3D Static Analysis

The low rock strength where the KV5 is excavated affects seriously the safety of the tomb both under static and seismic loading conditions. The PLAXIS 3D was used for

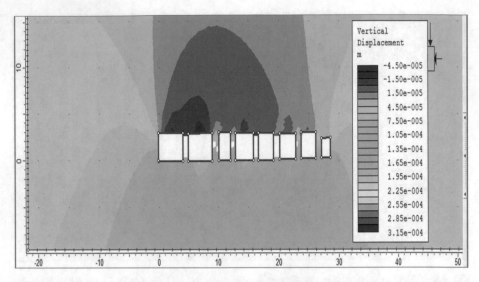

Fig. 15. Vertical displacement distribution through the rock pillars in Chamber 3. Examine 2D.

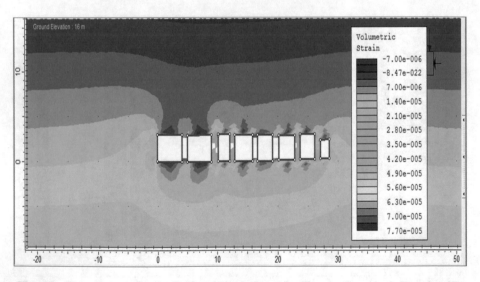

Fig. 16. Volumetric strain distribution through the rock pillars in Chamber 3. Examine 2D.

the 3-D numerical analysis of the central main Chamber with its sixteen supporting structural rock pillars.

A three-dimensional (3D) numerical model for the pillared chamber (3) (the largest chamber in Valley of the Kings (with its sixteen supporting rock pillars) and the large northern hall which are excavated in marl limestone deposit are constructed. The goal of the 3D examinations is to assess the pressure state in the columns considering the 3D geometry. The 3D impacts issue is considered on a fundamental designing

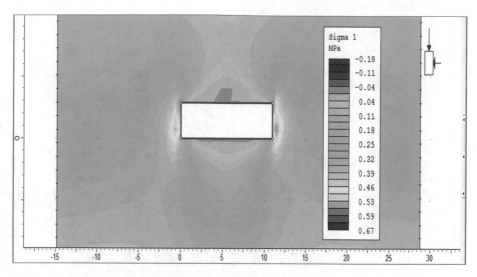

Fig. 17. Effective vertical stresses, the ceiling and sidewalls of northern Chamber. Examine 2D.

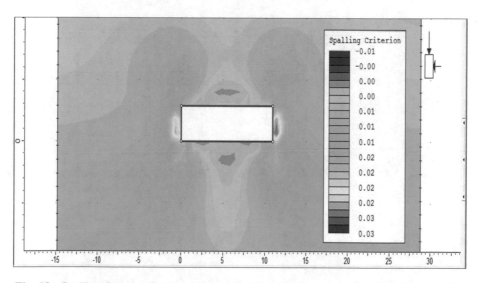

Fig. 18. Spalling Criterion for the ceiling and sidewalls of northern Chamber. Examine 2D.

methodology in the resulting segments. The different reproductions portrayed in this are directed utilizing the PLAXIS 3D code (PLAXIS 3D).

The results from the 3D static analysis which represented in Figs. 26, 27, 28 29 indicate that, the rock pillars in chamber 3 are under relatively high compression stresses. The calculated peak effective principal vertical compressive stresses on supporting rock pillars is 827.58 kN/m^2, the horizontal effective mean stresses 588.91 kN/m^2, the total displacement of the pillars 210.01×10^{-6} m, the vertical displacement

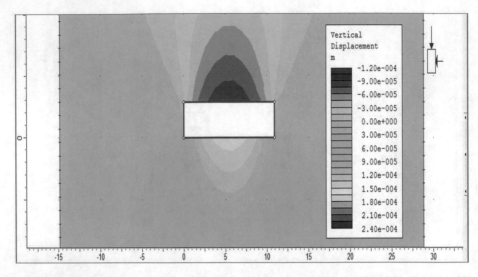

Fig. 19. Vertical displacement of the ceiling and sidewalls of northern Chamber. Examine 2D.

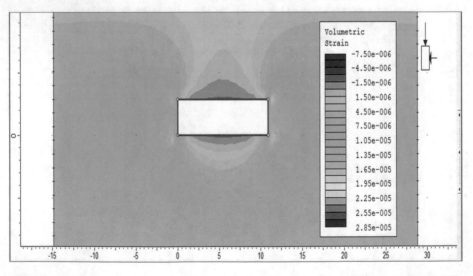

Fig. 20. Volumetric strains of the ceiling and sidewalls of northern Chamber. Examine 2D.

208.36×10^{-6} m, the horizontal displacement 32.94×10^{-6} m, the vertical incremental displacement 11.29×10^{-6} m, and the volumetric strain 3.62×10^{-3} %.

For the large northern chamber, the extreme effective mean stresses is 567.73 kN/m², the total displacement 475.95×10^{-6} m, the vertical displacement 475.59×10^{-6} m, the volumetric strain 12.42×10^{-3}%, the extreme volumetric strain incremental 1.38×10^{-3}%, and the horizontal displacement 53.60×10^{-6} m, Figs. 29, 30, 31, 32, 33, 34. Also the maximum vertical stress on the roofs and sidewall of Chamber (1) and

Fig. 21. Present state of the sixteen supporting rock pillars in the Chamber 3 which are structurally damaged. Vertical cracks due to the overloading and strength regression are obvious.

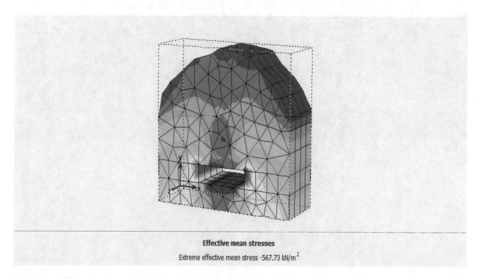

Effective mean stresses

Extreme effective mean stress -567.73 kN/m^2

Fig. 22. Effective mean stresses in the ceiling and sidewalls of northern Chamber. PLAXIS 3D.

Chamber (2) reached 688 kPa, on the separate wall between them, the 2D model did not calculate it, and the maximum vertical displacement of the ceiling reached 0.18×10^{-3}m, as shown in Fig. 22, 23, 24 25.

Figures 30 and 31 represent the analysis results of the large model which represents the complete east-west cross section of the tomb indicated that the stress distribution

144   S. Hemeda

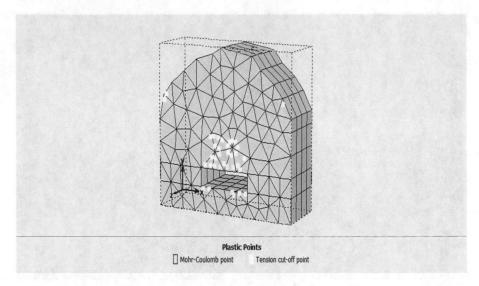

Fig. 23. Plastic Points in the ceiling and sidewalls of northern Chamber. PLAXIS 3D.

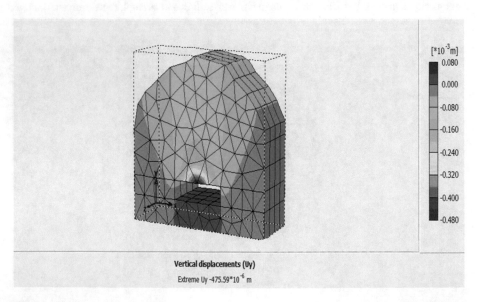

Fig. 24. Vertical displacement of the ceiling and sidewalls of northern Chamber. PLAXIS 3D.

and displacement values on the structural rock pillars in the Chamber (3) and Chamber (1) and (2) did not increase due to the excavation process extended behind the sixteen pillared Chamber 3 may it is due to the lowering of the ceiling level of these small burial chambers. Figure 30 represents the displacement progressive curve for the supporting rock pillars.

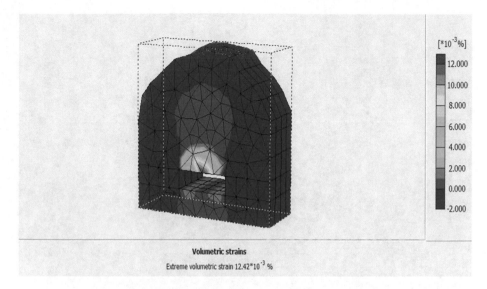

Fig. 25. Volumetric strains of the ceiling and sidewalls of northern Chamber. PLAXIS 3D.

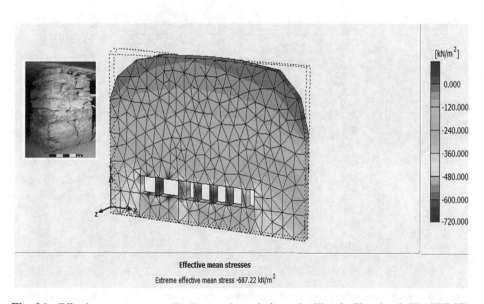

Fig. 26. Effective mean stresses distribution through the rock pillars in Chamber 3. PLAXIS 3D.

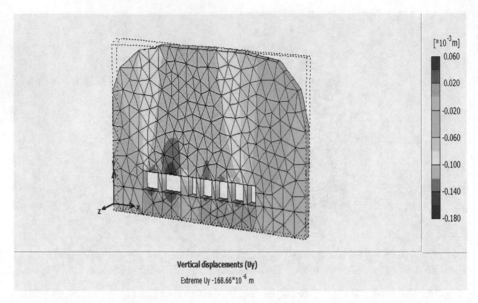

Fig. 27. Vertical displacements of the support rock pillars in Chamber 3. PLAXIS 3D.

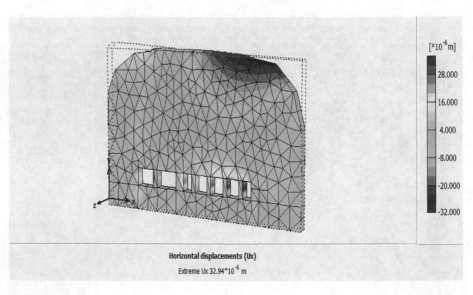

Fig. 28. Horizontal displacements of the support rock pillars in Chamber 3. PLAXIS 3D.

5.3   Evaluate the Safety Factor and Stress State in the Structural Support Pillars

It is demonstrated that induced stresses of significant magnitude and ambiguous distribution are to be expected in the supporting pillars. Multiple openings and

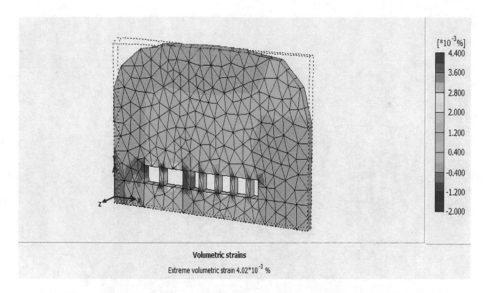

Volumetric strains

Extreme volumetric strain 4.02*10^{-3} %

Fig. 29. Volumetric strains of the support rock pillars in Chamber 3. PLAXIS 3D.

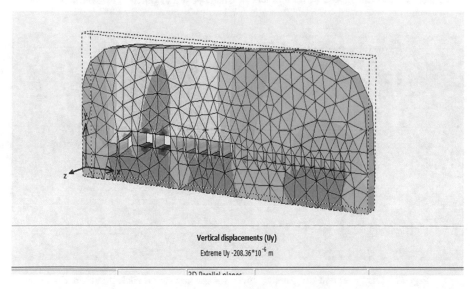

Vertical displacements (Uy)

Extreme Uy -208.36*10^{-6} m

Fig. 30. Vertical displacement of rock pillars in Chamber 3and others small burial chambers.

excavations designed on the basis of the average stress in the pillar σv^{-} given by the tributary area theory, as explained in Eq. (1) (Hemeda et al. 2010b).

$$\sigma v- = \frac{At}{AP}\sigma v \qquad (1)$$

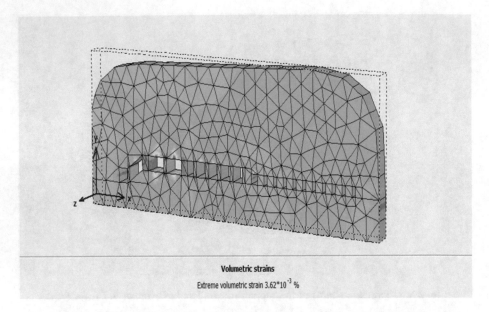

Fig. 31. Volumetric strains of rock pillars in Chamber 3 and others small burial chambers.

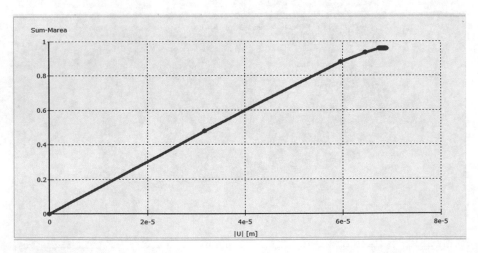

Fig. 32. Displacement progressive curve for the supported rock pillars in KV5.

Where,

- A_t is the area supported by the pillar
- A_p is the area of the pillar
- σ_v is the vertical stress at the level of the roof of the excavation (catacombs)

To evaluate the degree of safety of a pillar, we should be look at the better than expected column push σv with the column quality σp. The last isn't just the

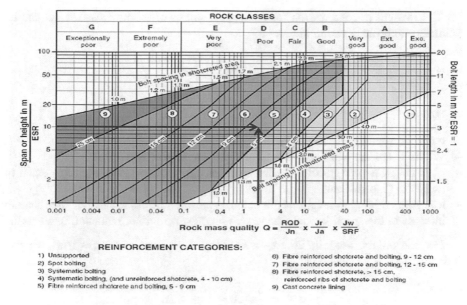

Fig. 33. Proposal support system for the (KV5) according to the Barton's Q-system.

Fig. 34. Temporarily unsuitable, expensive and eyesore support system for the KV5, installed recently by the Theban Mapping Project.

unconfined compressive quality of the material involving the column qu, in light of the fact that shape and estimate impacts present critical alterations from the breaking quality of unconfined compressive barrels.

The strength in compression for rectangular pillars of square cross section can be estimated from the equation number 2.

$$\sigma_p = \left\{0.875 + 0.250\frac{W}{H}\right\}\left\{\frac{h}{hcri}\right\}^{0.5}(q_u) \tag{2}$$

Where,

- σ_p is the strength of the pillar,
- W and H are the width and height of the pillar respectively,
- q_u is the UCS strength of the pillar material on cylinders with height (h) equal to twice the diameter and
- h_{crit} is the minimum height of the cubical specimen of pillar material such that an increase in the specimen dimension will produce no further reduction in strength

For the pillars, see Fig. 26, $\sigma_v = 700$ kPa, $A_t = 2$ m^2 and $A_p = 1$ m^2 we can derive:

$$\sigma v^- = \frac{2}{1}x700 = 1400\,KPa$$

The strength of the pillar σ_p can be estimated from the equation: For the pillar we have W = 1 m, H = 3 m. If we assume $h_{crit} = 0.2$ m and h = 1 m for $q_u = 900$ kPa, we have $\sigma_p = 1922$ kPa.

And the Factor of Safety $F.S = \frac{\sigma p}{\sigma v^-} = \frac{1922}{1400} = 1.37$ which very low and indicate to the dangerous and unsafe situation and losing of the structural function of these load bearing pillars. Hoek and Bray quote Salamon and Munro,s suggestion of acceptable safety factors >1.6. Such values may be adequate for the excavation stability, (Hemeda et al. 2010b). In the original study of Salamon and Munro this occurred between safety factors of 1.3–1.9 with the mean being 1.6. This value was recommended for the design of production pillars in South African bord and pillar workings (Salamon and Munro 1967).

$$\text{Also overstress state} = \frac{\sigma c}{\sigma v} = \frac{900\,KPa}{700\,kPa} = 1.28\,MPa \tag{3}$$

The tributary theory is based on average pillar stresses and derived stress value is generally close to the averages predicted by PLAXIS 3D. On other hand, the over-loading of geostatic loading due to the overburden strata on the supporting rock pillars is obvious and it induced critical vertical cracks in these pillars. also some sections have an overriding influence on the pillar stability, referring to Eq. (3), particularly in terms of long-term creep effects and associated strength loss or thinning-out of the effective load bearing pillars and section (Hemeda 2008).

6   Design of Structural Supporting System

6.1   The First Option, Which Depend on the RMR

Rock Mass Rating system is based on combination of six parameters = Intact Rock Strength, RQD, Joint Spacing, Joint Conditions, Groundwater and Adjustment factor.

The first option depends on the Bieniawski's RMR (Bieniawski 1989) (Rock Mass Rating System) calculation, where the strength of intact rock is 900 kPa (with rate 1), the RQD is 50 (with rate 13), the spacing of joints less than 60 mm (with rate 5), the conditions of discontinuities is Separation 1–5 mm with Continuous joints (with rate 10)and the ground water conditions are completely dry (with rate 15), for the adjustment for joint orientation is −5 then The RMR of the Sons of Ramses II tomb is (39) which classified as poor rock with high stresses. As shown in Table 4.

Table 4. RMR value for the KV5 is determined as follow.

Item	Value	Rating
Uniaxial compressive strength	900 kPa	1
RQD	50	13
Spacing of discontinuities	<60 mm	5
	Separation 1–5 mm	10
	Conditions of discontinuities	Continuous joints
Ground water	Completely dry	15
Adjustment for Joint Orientation		−5
Total RMR		39 Poor rock

According to the RMR value, the design of the support system for the pillars and whole KV5 can include Systematic bolts 4–5 m long, spaced 1–1.5 m in Crown and walls with wiremesh. 100–150 mm in Crown and100 mm insides with Light to medium ribs Spaced 1.5 m where required.

6.2   The Second Option, Which Depend on the Q-System

The 2nd option depends on the Barton's Q- system or the rock tunneling quality index of the rock mass where the tomb is excavated.

The Q-system of Barton et al. (1974), Barton (1988) expresses the quality of the rock mass in the so-called Q-value. The Q-value is determined as follows, Eq. 4:

From the Q- system parameters which include the RQD is 50, Jn with value 4, Jr with value 3, Ja with value 1, Jw with value 1.

$$Q = \frac{RQD}{Jn} x \frac{Jr}{Ja} x \frac{Jw}{SRF} \tag{4}$$

For a depth below surface of 17 m, the overburden stress will be approximately $17 \ m^2 \times 21 \ kN/ m^3 = 396$ kPa. The major principal stress σ1 is $2 \times 396 = 792$ kPa.

Given the uniaxial compressive strength of the supporting rock pillars is approximate 900 kPa, this gives a ratio of $\frac{\sigma c}{\sigma 1} = \frac{900}{792} = 1.136 < 2.5$ which refer to a high stressed poor rock with SRF 20, then Q or rock mass quality value is 1.87 (poor rock according to the Q-system) as shown in Table 5.

Table 5. Rock Tunneling quality index, Q-system determined as follow.

Parameter	Description	Value
RQD	Rock quality designation	50
Jn	Joint number	4
Jr	Joint roughness	3
Ja	Joint alteration	1
Jw	Joint water reduction factor	1
SRF	Stress reduction factor	1.13
Total Q-System		1.87 Poor rock

For an excavation span (the width of the pillared chamber 3) of 15.6 m, the equivalent diameter, De = 15.46/1.6 = 9.66, where the ESR or the permanent opening is 1.6 and the width of the pillared chamber 3 is 15.46 m.

The value of De of 9.66 and value of Q of 1.87 places the tomb of sons of Ramses II in category (5) which require Fiber Reinforced Shotcrete and bolting 5–9 cm. Length of rockbolts with L = 2 + (0.15B/ESR) and Maximum span (unsupported) = 2 ESR × Q^{04}. As shown in Fig. 33. The strength properties of FRPs collectively make up one of the primary reasons for which select them in the strengthening and seismic retrofitting. A material's strength is governed by its ability to sustain a load without excessive deformation or failure. Also it is recommended to use the CFRP, also nowadays we can use the advanced or Nano CFRP because of its good mechanical properties in particularly the compressive and tensile strength.

6.3   The Third Option

The third option for the permanent support for this complex kind of underground structures could be designed as presented in Fig. 35, where the rock bolts with 4–9 cm and prestressed anchors or micro piles with 100 mm Diameter for the permanent support system for the rock pillars and sidewalls of the KV5.

7   Discussion of Numerical, Laboratory Analysis and the Field Observation

The rock mass which the sons of Ramses II tomb is excavated can be classified as moderately to extensively jointed or fractured rock contains joints and hair cracks, but the blocks between joints are locally grown together or so intimately interlocked that

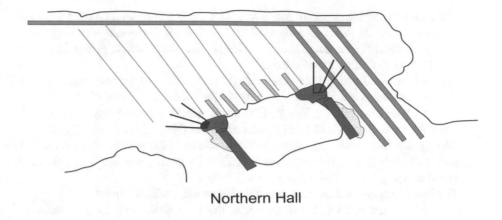

Northern Hall

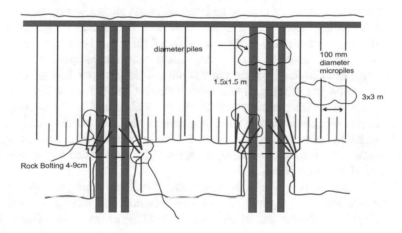

Chamber 3

Fig. 35. Design of rockbolts, prestressed anchors and micropiles for the permanent support system for the rock pillars and sidewalls of the KV5.

vertical walls do not require lateral support. In rocks of this type, both spalling and popping conditions may be encountered.

It is notice that In brittle rock, high stress conditions may lead to rock bursting (the sudden release of stored strain energy) bursts manifest themselves through sudden, as shown in Fig. 9.

Analysis and interpretation of the numerical and laboratory results and the field observations led to the following findings:-

1. Most of the Royal Tombs in the Valley of the Kings were excavated into the marls

of the middle and lower part of Member I. Thin-sections prepared on the limestone samples where the KV5 is excavated, refers that the limestone is fine-grained calcite, embedded in a micritic matrix rich in amorphous silica, fossils like Foraminifera and large grains of quartz.

2. Presence of swelling-type clay minerals (Montmorillonite) in some rocks of the Thebes Formation but, more importantly, in the underlying rocks of Esna formation. (Clay layers swelling). The XRD analysis indicated that the major contents of Esna shale are quartz (SiO_2) and Montmorillonite ($Na_{0.2}$ $Ca_{0.1}$ Al_2 Si_4O_{10} $(OH)_2$. $(H_2O)_{10}$, the minor contents include the Kaolinite and Illite with Calcite traces. The bulk unit weight of the Esna shale is 1.79–1.86 g/cm^3, and the uniaxial compressive strength is 4.22–4.43 kg/ cm^2.

3. The index properties show that the shale layers are medium expansive.

4. Anhydrite found in abundant quantities in the Esna Shale, may be a factor contributing to swelling of the Esna Shale.

5. Rock slope deformations (spreading) of the Thebes limestone blocks caused by volume changes in the underlying Esna shale. As shown in Fig. 6.

6. The lowermost unit of the Thebes Formation. Nonetheless, KV5 enter into the fundamental interbedded shale and marls of the Esna Formation. Every one of them indicates extreme, irreversible shake structure weakening starting from swelling and shrinkage. Water and flotsam and jetsam from the past and late blaze surges impactsly affected divider embellishment of the highest chambers and on columns and divider structure in the chambers (1, 2 and 3). Noteworthy flooding since the disclosure of the tomb has caused real pulverization of dividers and columns by rehashed swelling and shrinkage of the shale. Besides, quickened moistness changes in the course of recent years have added to expanding weakening of the stone structure.

7. The dislodging of shake units before the tomb uncovering brought about inexhaustible shake joints, which can be re-initiated amid seismic tremors or other quick pressure discharges, for example, by swelling of the shale. At the point when water enters the tombs, it comes into contact with the shale at the lower chambers, and causes swelling, breaking and basic disappointments in the floors, dividers, and columns.

8. Gravity rock falls and sliding of rock features along inclined discontinuities at the surrounding area.

9. Extensive jointing (rock discontinuities) present in the rock at tomb depth.

10. The overloading of geostatic loading due to the overburden strata on the supporting rock pillars is obvious and it induced critical vertical cracks in these pillars also some sections have an overriding influence on the pillar stability, particularly in terms of long-term creep effects and associated strength loss or thinning-out of the effective load bearing pillars and section, as shown in Fig. 21.

11. Rock detachment and falls from the ceiling. As shown in Fig. 9.

12. Shape and measures deformations of the tombs or some sections of them.

13. Detachment and falls of renders with its wall paintings.

14. Intensive weathering and erosion of lower parts of the structural elements in particularly the supporting rock pillars in Chamber 3. The main structural deficiency attributed to the impact of flash floods in the past and few years ago. As shown in Fig. 8.
15. Physical, mechanical and chemical changes in the construction materials. The strength reduction is obvious and the UCS reached in some critical sections and supporting rock pillars to less than 1 MPa. Those secondary fossils content, due basically on shells about foraminifers and a portion mollusks, provide for climb on structural heterogeneity, which reflected in the variability of the mechanical properties What's more in the poor reproducible of the test results (Bukovansky et al. 1997).
16. Nearness of huge, vertical, open cracks at the surface of slopes on the two sides of the valley. These cracks can be followed both in the valley partitions where the tombs are found. The cracks are effortlessly unmistakable in the majority of the region. The starting point of these breaks has never been translated, albeit a few geologists viewed them as flaws.
17. The numerical analysis results indicated that the safety factor of the structural support rock pillars in chamber 3 is very low in order of 1.37 and the overstress state is 1.28 MPa.

Remedial and retrofitting policies and techniques, static monitoring and control systems which are necessary for the strengthening and stability enhancement of the tomb, where the rock mass classification indicated the rock mass where the KV5 is excavated is poor rock, with RMR 39 and Q value 1.87.

Systematic bolts with 4–5 m long, spaced 1–1.5 m in Crown and walls with wiremesh. 100–150 mm in Crown and 100 mm insides with Light to medium ribs Spaced 1.5 m where required for strengthening retrofitting of the KV5. Also it is recommended to use Fiber Reinforced Shotcrete and bolting 5–9 cm. Length of rock bolts $L = 2 + (0.15B/ESR)$ and maximum span (unsupported) $= 2$ ESR X Q^{04}; also nowadays we can use the advanced or nano carbon tubes because of its advanced physical and mechanical properties in particularly the compressive and shear strength. The third proposal is the installation of rock bolts with 4–9 cm and prestressed anchors or micro piles with 100 mm Diameter for the permanent support system for the rock pillars and sidewalls of the KV5.

8 Conclusions

We can state that most of what we can call now geotechnical problems was faced in the Valley of Kings where most of the large important subterranean decorated tombs of the pharaohs like sons of Ramses II tomb KV5 are found. This case study illustrates how the quantification of various variables permits an understanding of the problems facing a site and also suggests possible solutions.

In conclusion the detailed engineering analysis of the sons of Ramses II tomb KV5 at Luxor, Egypt proved that these unique monuments present low safety factors of the rock pillars which are structurally damaged, where the factor of safety F.S is about

1.37, (note that the acceptable safety factor for the underground structures is >1.6 in static state). Also the overstress state of the surrounding rocks is beyond the elastic regime (limit of domain), and all the rock pillars structural supports are subjected to high vertical compressive stresses. Many instability problems for static and dynamic loading were recorded and analyzed. Consequently a well-focused strengthening and retrofitting program is deemed necessary.

a. Acknowledgements

Not applicable

b. Availability of data and materials

Not applicable

c. Competing interests

The author declare that he has no competing interests

d. Funding

The author confirms that he is not currently in receipt of any research funding relating to the research presented in this manuscript

e. Availability of Supporting Data

"Data sharing not applicable to this article as no datasets were generated or analyzed during the current study"

f. List of Symbols and Abbreviations

Symbols

At	is the area supported by the pillar
Ap	is the area of the pillar
σv	is the vertical stress at the level of the roof of the excavation (KV5)
σp	is the strength of the pillar,
W and **H**	are the width and height of the pillar respectively,
qu	is the UCS strength of the pillar material on cylinders with height (h) equal to twice the diameter
hcrit	is the minimum height of the cubical specimen of pillar material such that an increase in the specimen dimension will produce no further reduction in strength.
β	Angle between the normal to the fracture plane and the horizontal plane
φ	Friction angle of the fracture
x τ	Shear stress in resin annulus
σ b	Applied stress
α	Decay coefficient 1/in which depends on the stiffness of the system
β	Reduction coefficient of dilation angle
σ c	Uniaxial compressive strength of rock
Aj	Joint area
φ b	basic joint friction angle

Ds	Rib spacing
U	The shear displacement at each step of loading
c	Cohesion between block joints
σ n	Normal force
b u	Shear displacement
N p	Normal force at failure
Qp	Shear force at failure
M D	Bending moment at yield limit
M p	Bending moment at plastic limit
Ei	Modulus of elasticity of intact rock
Qcf	Shear force
Lcp	Reaction length
v	Poison ration of rock mass
Po	In situ stress

Abbreviations

JRC	Joint roughness coefficient
JCS	Joint compressive strength
RMR	rock mass rating
RQD	rock mass Designation
Q	rock mass quality
Jn	joint set number
Jr	joint roughness number
Ja	joint alteration number
Jw	joint water reduction factor
SRF	stress reduction factor

References

Barton, N.R., Lien, R., Lunda, J.: Engineering classification of rock masses for the design of tunnel support. Rock Mech. Rock Eng. Springer. **6**(4), 189–236 (1974)

Barton, N.R.: Rock mass classification and tunnel reinforcement selection using the Q-system. In Kirkaldie, L. rock classification system for engineering purposes: ASTM special technical publication 984.1. ASTM International, pp. 59–88 (1988)

Bieniawski, Z.T.: Engineering Rock Mass Classifications. Wiley, New York (1989)

Brown, E.T.: Risk assessment and management in underground rock engineering—an overview. J. Rock Mech. Geotech. Eng. **4**(3), 193–204 (2012)

Bukovansky, M., Richard, D.P., Week, K.R.: Influence of slope deformations on the tombs in the valley of the kings, Egypt. Proc. Int. Symp. Eng. Geol. Env. **3**(1997), 3077–3080 (1997)

Deere, D.U., Varde, O.A.: General report, engineering geological problems related to foundations and excavations in weak rocks. In: Proceedings of the 5th International Association of Engineering Geology Congress, vol. 4, pp. 2503–2518 (1990)

Examine 2D v.8.0 Program from Rocscience (2D stress analysis for underground excavations software) Examine 2d. http://www.cesdb.com

Hemeda, S.: An integrated approach for the pathology assessment and protection of underground monuments in seismic regions. Application on some Greek-Roman monuments in Alexandria, Egypt. Ph.D Thesis, Civil Engineering Department, Aristotle University of Thessaloniki, Greece (2008)

Hemeda, S., Pitlakis, K.: Serapeum temple and the ancient annex daughter library in Alexandria, Egypt: geotechnical–geophysical investigations and stability analysis under static and seismic conditions. Eng. Geol. **113**, 33–43 (2010)

Hemeda, S., Pitilakis, K., Bakasis, E.: Three – dimensional stability analysis of the central rotunda of the catacombs of Kom El-Shoqafa, Alexandria, Egypt. In: 5th International Conference In Geotechnical Earthquake Engineering and Soil Dynamics. San Diego, California, USA, 24–29 May 2010 (2010b)

Dunn, J. (2014) The Geography and Geology of the Valley of the Kings on the West Bank at Thebes, London

PLAXIS 3D SOFTWARE. INFO@ PLAXIS.COM

PLAXIS Manual, Finite element code for soil and rock analysis. Published and distributed by A A Balkema Publishers, Nederland's Comput. Geotech. **32**(5), 326–339

RocLab 1.0 Software program for determine rock mass strength from Rocscience

Salamon, M.D.G., Munro, A.H.: A Study of the strength of coal pillars. J. S. Afr. Inst. Min. Metall. **68**(2), 55–67 (1967)

Siliotti, A.: Guide to the Valley of the Kings. Barnes & Noble Books, Newport News (1997)

Theban Mapping Project, [Regular reports on the work in KV 5 appear on the Project's website. www.thebanmappingproject.com]

Weeks, K.R.: The Theban Mapping Project: Report of the 1994 Field Season. Theban Mapping Project, Cairo (1994)

Weeks, K.R.: The work of the Theban Mapping Project and the protection of the valley of the kings. In: Wilkinson, R. (ed.) Valley of the Sun Kings: New Expeditions in the Tombs of the Pharaohs. University of Arizona Egyptian Expedition, Tucson (1995)

Weeks, K.R.: The Lost Tomb. William Morrow and Company, New York (1998)

Weeks, K.R.: Atlas of the Valley of the Kings. Publications of the Theban Mapping Project, American University in Cairo Press, Cairo (2000)

Weeks, K.R.: KV5: A Preliminary Report on the Excavation of the Tomb of the Sons of Ramesses II in the Valley of the Kings. The American University in Cairo Press (2006)

Wüst, R., McLane, J.: Rock deterioration in the Royal Tomb of Seti I, valley of the kings, Luxor, Egypt. Eng. Geol. **58**, 163–190 (2000)

Recovery Process of the Water Distribution System After a Seismic Event

Yolanda Alberto-Hernandez[(✉)]

Graduate School of Engineering, The University of Tokyo, Tokyo, Japan
h-yolanda@t-adm.t.u-tokyo.ac.jp

Abstract. The water distribution network of a city is one of the most critical infrastructures. The damage it sustains during earthquakes and the ease of the restoration process define the resiliency of the system and impacts other critical lifelines. This paper presents a summary of the damage in two cities of Japan after the 2011 Tohoku Earthquake and the 2016 Kumamoto Earthquake. It focuses on the recovery process, exhibits logistical inter-dependence with other lifelines and provides information on the post-disaster data collection. An example of the actions required for recovery is given. Conclusions provide advice on the most important factors to consider for reducing damage in future events, and on the appropriate way to collect post-disaster information that can be used for decision support systems.

Keywords: Seismic vulnerability · Lifelines · Water pipelines leakage
Water distribution pipelines seismic analysis · Network analysis
Recovery · Decision support system

1 Introduction

On March 11, 2011, a M_w 9.0 earthquake stroke the east coast of Japan, triggering a tsunami and several impacts, that included the nuclear accident in Fukushima. This event, one of the five most powerful earthquakes in the world since 1900, was followed by two aftershocks inducing additional damage of M_w 7.4 and M_w 7.7, 15 and 30 min after the first event (Koshimura et al. 2014). The Great East Japan Earthquake Disaster, as called by Cabinet decision, is the worst event of modern times in Japan. In spite of having enough provisions regarding seismic design and evacuation, the unexpected magnitude of the tsunami startled the affected prefectures where several people failed to evacuate due to early erroneous predictions, buildings selected as safe points were washed away and structures that survived the earthquake experienced severe damage. The Geospatial Information Authority (Government of Japan 2011) estimated an affected area of 561 km^2 by the combined effect of earthquake and tsunami. Damage from the 2011 Great East Japan Earthquake Disaster can be classified from its source as earthquake-induced or tsunami-induced. As for the performance of lifelines, there was partial shutdown of the electrical service that was restored within the following next five days. The gas network was stopped in local areas to avoid fires and it was kept on

© Springer Nature Switzerland AG 2019
M. Badr and A. Lotfy (Eds.): GeoMEast 2018, SUCI, pp. 159–168, 2019.
https://doi.org/10.1007/978-3-030-01884-9_11

hold to help the recovery of the water network. Transportation networks were affected by ground shaking, including landslides, liquefaction and lateral displacement. The mobile network was affected by the power outages, some of the main exchanges connections were disrupted and one cell tower collapsed. Out of these phenomena, liquefaction of reclaimed lands, young alluvium deposits or river levees, instability of embankment and slope failure were the main problems.

On April 2016, two earthquakes hit the area of Kumamoto, on the south part of the Kyushu island of Japan. The first fore shock, on the 14th a M_w 6.0 hit the area and on the 16th a M_w 7.0 earthquake affected the Kumamoto Prefecture. The cities of Kumamoto and Mashiki, reported 70 casualties and 1,820 injured people. Residential properties experienced total and partial collapse (8,044 and 24,274, respectively), and about 118,222 houses experienced a certain amount of damage (Tang and Eidinger 2017).

These two earthquakes affected the lifelines of several cities, including the water distribution network. This paper presents a report of damage and recovery of the water system and presents a neurofuzzy decision support system approach for identifying the key factors for protection and recovery, but also for post-earthquake data collection.

Methodology followed: archival study, observations and semi-structured interviews for collecting data, to find the important factors, they were evaluated and analyzed by tracking the decision-making process, examining resultant consequences and foreseeing onward challenges

2   Urayasu City

There were around 2,105,091 houses with water outages as of March 11, 2011 in 187 municipalities, 616, 480 houses in Miyagi prefecture, 444,288 houses in Ibaraki prefecture and 373,069 houses in Chiba prefecture according to the Ministry of Health, Labour and Welfare, the percentage of service reconnection is shown in Fig. 1. Regarding sewage damage, there were reports of failures in 120 treatment facilities, 112 pump stations, 1061 km pipelines and 20730 manholes according to the Ministry of Land, Infrastructure, Transport and Tourism. The cause was a combination of liquefaction, strong ground motion and tsunami waves (Naba et al. 2012).

There are different geotechnical hazards for the water system, depending on the component. The caption structures can be affected by landslides or slope failures that may damage the caption structure or expose the pipelines. The pipelines must comply with the required technical specifications of installation, be at the proper depth and have a compacted backfill, this can also be affected by climate changes and breakage due to ground vibration, pipelines can also report leakage which leads to major failures when there is an earthquake.

After the earthquake, severe liquefaction-induced damage was observed in Tokyo Bay, where reclamation of the coastline started around 1600, from Sumida to Yokohama and expanded to Kanagawa and Chiba Prefectures around the 50's. These areas were affected before by other seismic events. On December 17, 1987, a M_w 6.7

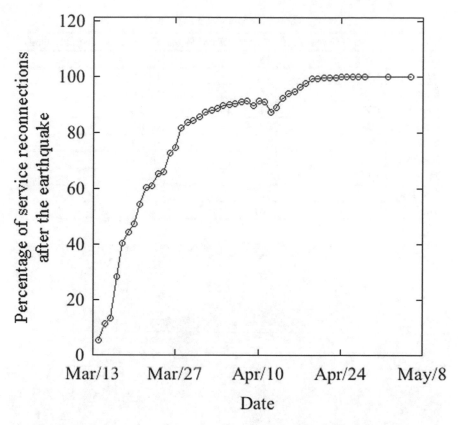

Fig. 1. Percentage of service reconnection after 2.2 millions of outages in Japan

Earthquake hit eastern Chiba Prefecture (Chibaken Toho-oki Earthquake) causing liquefaction in many areas of the city, including Kairaku 1-chome, Mihama 3-chome and Irifune 4-chome. Therefore, some zones that experienced liquefaction in 1987 liquefied again. On March 2011, boiled sand was observed in the reclaimed cities of Urayasu, Ichikawa, Narashino, Odaiba, Shinonome, Tatsumi, Toyosu, Seishin, Yokohama, Kawasaki, Kizarasu and Chiba (Yasuda and Ishikawa 2012). Table 1 shows the number of leaks in the cities of Chiba Prefecture. However, the most devastating effects were found on lands reclaimed after the 70's where differential settlement and lateral displacement affected private houses, roads, sea walls, pipelines and other structures. Amid those areas, Urayasu City was the most affected, where more than 9,000 private properties had detriment (Tokimatsu and Katsumata 2012). The location of breaks in Urayasu City are displayed in Fig. 2.

Table 1. Damage rate by city

City	Water pipelines (km)	Number of breakages	Damage rate (Breakage/km)
Chiba	2,072	45	0.017
Ichihara	1,141	1	0.001
Funabashi	1,397	43	0.031
Narashino	178	24	0.135
Kamagaya	318	0	0.000
Narita	101	3	0.030
Ichikawa	1,027	10	0.019
Matsudo	1,143	5	0.004
Urayasu	300	321	1.070
Total	8,307	461	0.055

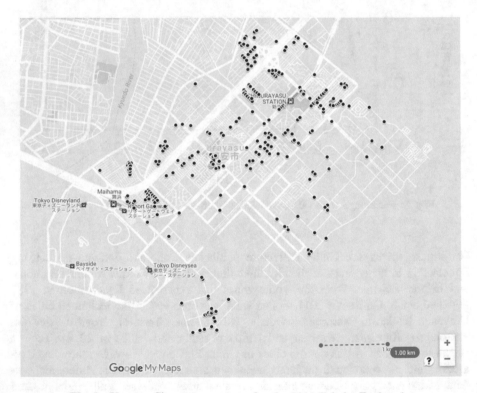

Fig. 2. Urayasu City water outages after the 2011 Tohoku Earthquake

The damage rate in Urayasu City is shown in Tables 2 and 3 by material and diameter, respectively.

The restoration curve is shown in Fig. 3, where it is observed that performance decreased to 0 immediately after the earthquake due to the necessary outages done to check the system and then went up to 85% after 4 days. In this case, the use of tank trucks delivering water, was one of the main strategies to restore the service.

Table 2. Damage rate by material in Urayasu City

Material	Length of water pipe (km)	Number bof breaks	Damage rate (breaks/km)
KDP	313	13	0.042
TDP	5,708	285	0.050
ADP	1,218	151	0.124
HIVP	252	7	0.028
Others	817	5	0.006
Total	8,307	461	0.055

Table 3. Damage rate by diameter in Urayasu City

Conducting water pipe	Diameter	$\phi50$	$\phi75$	$\phi100$	$\phi150$	$\phi200$	$\phi300$	$\phi350$–1500	Total
	Number of water leaks	9.0	90.0	113.0	106.0	76.0	48.0	20.0	462
	Percentage	1.9	19.5	24.5	22.9	16.5	10.4	4.3	100
Water supply pipe	Diameter	$\phi13$	$\phi20$	$\phi25$	$\phi40$	$\phi50$	$\phi75$	$\phi100$–200	Total
	Number of water leaks	4.0	141.0	25.0	56.0	35.0	12.0	9.0	282
	Percentage	1.4	50.0	8.9	19.9	12.4	4.3	3.2	100

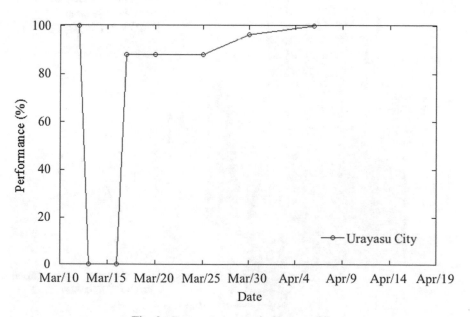

Fig. 3. Restoration curve in Urayasu City

3 Kumamoto City

On April 14 and 16, 2016, two earthquakes occurred at the southern island of Kyushu, in Kumamoto Prefecture. The first, a foreshock of M_w 6.0, occurred at 21:26 local time, with the epicenter 32.849°N 230.635°E, at a depth of 10 km. The main shock M_w 7.1, stroke the area 28 h later, at 1:25 local time with epicenter at 32.782°N 130.726°E at a depth of 10 km. There were 69 casualties, 364 seriously injured, 8044 collapsed houses, 24,274 partially collapsed and 118, 222 with some damage.

Lifelines were heavily damaged both due to high inertial shaking as well as due to PGDs. There was few liquefaction due to the short duration and PGA < 0.2 g on alluvial soils. Liquefaction occurred further to the west, along and within flood plains.

The water system of Kumamoto city is composed of 113 wells, 19 pump stations, 61 distribution tanks with a total storage of 58 million gallons; 3,366 km of pipe, ranging from 3 to 60 inches-diameter. Pipe materials are 577 km of earthquake resistant ductile iron pipe (ERDIP), 1,8882 km of ductile iron pipe (DIP), 73 km of welded steel pipe, 424 km of PVC pipe, 136 km of High density polyethylene pipe (HDPE), 81 km of cast iron pipe (CIP) or other. The total water supply capacity of 84 MGD (316, 116 m3/day), an average day demand of 58 MGD (218, 171 m3/day) and a population served of 740,000 people. Since 2005, the prefecture has only installed seismic resistant water

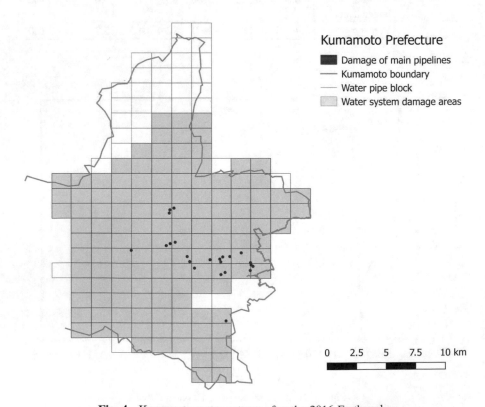

Fig. 4. Kumamoto water outages after the 2016 Earthquake

pipe, the earliest installed was done in 1979 and as of 2014, 75% of all 14-inch diameter and larger (trunk lines) are "seismic resistant" pipe, 22% of all pipe are "seismic" pipe.

Out of the different cities existing in Kumamoto Prefecture, there was a loss of water supply in 445,000 households, having 327,000 in Kumamoto City, 31,000 in Ozu Town, 11,000 in Mashiki Town, 10,000 in Aso Town and 50,000 in smaller communities. The prefectures nearby, Oita and Kiyazaki, were also affected with 10,000 and 2,800 water outages, respectively. The city budgeted 410 million USD for the period of 2010–2028 to integrate the smaller nearby water systems into the larger City water system, plus 310 million USD to make seismic improvements to the existing system. The affected areas are displayed in Fig. 4.

In the major soft ground or in the liquefaction-prone areas, all trunk pipes are welded steel pipes. Within these liquefaction zones, there were no known failures of pipes, but on the limit of the liquefaction area, where there is a transition in materials, there were reports of failures. By June 2016, the repairs statistics in the city were: 125 trunk pipe repairs; 107 distribution pipe repairs; 1,574 service line repairs, 1,638 customer side investigation and repairs if needed; 17 wells with damage, including pipes 5, buildings 12, electrical instrumentation 2, pump 1, fences 2 and ground settlements 12; 1 reservoir with damage; 2 pump stations with damage, including pipes 3, building 2, ground settlement 2; 7 tanks with damage, including pipes 3, instrumentation 1, fence 1, ground settlement 3. The restoration curves for both Kumamoto Prefecture and Kumamoto City are shown in Fig. 5.

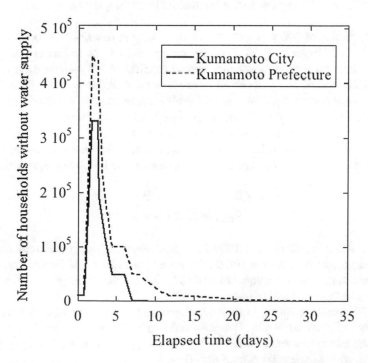

Fig. 5. Restoration curve un Kumamoto Prefecture and Kumamoto City

In Kumamoto City a large number of water leaks was caused by the damage to the valve plug body and the flange part, and some countermeasures are required for maintenance renewal in the future. After the earthquake, the turbidity generated caused a large interruption on a wide area that left the population without service. There were hundreds of potable water pipe breaks on non-seismically design buried pipes (older materials), thousands of repairs for service lateral, and thousands of repairs for service lateral. 445,000 households lost water supply, 327,000 Kumamoto City and 11,000 Mashiki.

There was a good response of the most modern pipelines. There were no failures of welded steel pipes on liquefiable areas but on the transition area there were failures. In total, there were 125 trunk pipe repairs, 107 distribution pipe repairs, 1574 service line repairs, 1638 "customer side" investigations, 17 wells with damage (including 5 pipes, 12 buildings, 2 electrical instrumentation, 1 pump, 2 fences, 12 ground settlements); 1 reservoir with damage, 2 pump stations with damage (including 3 pipes, 1 building and 2 ground settlement); 7 tanks with damage (including 3 pipes, 1 instrumentation, 1 fence and 3 ground settlements) (Tang and Eidinger 2017).

4 Decision Making Process

There are three main stages after an earthquake: emergency response, damage assessment and restoration. The selection of the activities that need to be carried out in each stage is one of the most important factors in developing an adequate emergency plan.

The particular recovery process for Urayasu City is described in the flow chart of Fig. 6 (City of Urayasu 2017). Primary survey consists of sediment dredging and manhole visual inspection. Secondary survey consists of TV camera survey, detailed survey by inspectors, and assessment of abnormal conditions.

The survey involved all the parties involved, government, residents, researchers, companies, etc. Collection of data regarding breaks and failures was essential to consider the strategies that were necessary for the recovery.

According to American Lifelines Alliance, ALA, (2001) the seismic vulnerability of buried pipelines and their repair rate are given by the following equations:

$$RR_{PGV} = K_1 \bullet a \bullet PGV$$
$$RR_{PGD} = K_2 \bullet b \bullet PGD^c$$

Where RR is repair rate per 1000 ft of pipe length, peak ground velocity (PGV), permanent ground deformation (PGD), in inches per second and inches, respectively. The constants, a, b and c are equal to 0.00187, 1.06 and 0.319, respectively. K_1 and K_2 depend on the material of the pipe.

The risk factors involved in seismic mitigation are age of pipe, diameter, material and number of previous breaks. Diameter and number of previous breaks are the most influencing factors in areas affected by liquefaction, given that most of the pipes have been replaced with seismic-resistant materials.

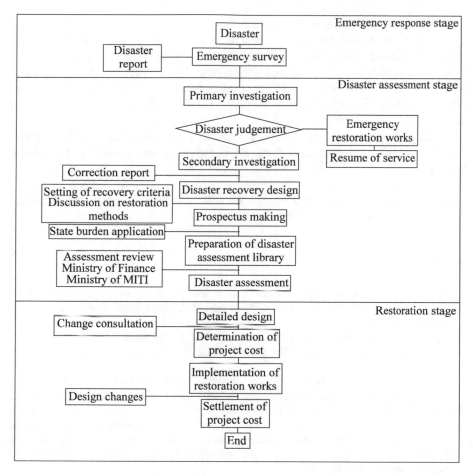

Fig. 6. Flow of disaster restoration in Urayasu City

Considering these factors, in the flow chart is observed how the preparation of an assessment library, which included mapping the data using GIS was a required point before preparing a restoration project. This data can be later used to make network seismic analysis to identify the most vulnerable points and can be added to decision support systems for the Waterworks Bureau to define where to make repairs or replacements.

5 Conclusions

Water distribution system is one of the main assets of a city along with other lifelines and must be quickly restored after an earthquake. In the cities of Urayasu and Kumamoto, the liquefaction phenomena affected greatly the water pipelines, causing water outages for long periods of time. Replacement and repair of pipelines was necessary in order to resume the service. In order to increase the resilience of this system, the use of artificial learning systems could be helpful to identify the most vulnerable areas in the network and determine if it is necessary to make repairs or replacements. However, this requires a very large data set that includes detail information of the characteristics of the pipelines and their history. To achieve this, recollection of data after an earthquake should be done carefully to determine the damage state and pipeline conditions.

In the future, the author would like to build a neurofuzzy system that is integrated with a geographic information system and includes the analysis of the geotechnical hazards.

Acknowledgements. The author thanks the Water Bureau Office of Urayasu and Kumamoto cities for the information provided for this research.

References

American Lifelines Alliance (ALA): Seismic fragility formulation for water systems, 104 (2001)

City of Urayasu: 2011 Tohoku earthquake. Report of Urayasu City, Chiba Prefecture (in Japanese) (2017)

Government of Japan: Regarding the 2011 off the Pacific coast of Tohoku earthquake (The great east Japan earthquake disaster). The Cabinet Office Headquarters for Emergency Disaster (2011)

Koshimura, S., Hayashi, S., Gokon, H.: The impact of the 2011 Tohoku earthquake tsunami disaster and implications to the reconstruction. Soils Found. Elsevier **54**(4), 560–572 (2014)

Naba, S., Tsukiji, T., Shoji, G., Nagata, S.: Damage assessment on water supply system and sewerage system at the 2011 off the Pacific coast of Tohoku earthquake - case study for the data at Ibaraki and Chiba prefectures. In: Proceedings of the International Symposium on Engineering Lessons Learned from the 2011 Great East Japan Earthquake, 1–4 March 2012, Tokyo, Japan, 1487–1495 (2012)

Tang, A. K., Eidinger, J.M.: Kumamoto, Kyushu, Japan earthquakes of Mw 6.0 and Mw 7.0 April, 2016. Lifeline performance. Technical Council on Lifeline Earthquake Engineering (2017)

Tokimatsu, K., Katsumata, K.: Liquefaction induced damage to buildings in Urayasu City during the 2011 Tohoku Pacific earthquake. In: Proceedings of the International Symposium on Engineering Lessons Learned from the 2011 Great East Japan Earthquake, 1–4 March 2012, Tokyo, Japan, pp. 665–674 (2012)

Yasuda, S., Ishikawa, K.: Several features of liquefaction-induced damage to houses and buried lifelines during the 2011 great east Japan. In: Proceedings of the International Symposium on Engineering Lessons Learned from the 2011 Great East Japan Earthquake, 1–4 March 2012, Tokyo, Japan, pp. 825–836 (2012)

Mechanically Stabilized Granular Layers - An Effective Solution for Tunnel Projects

Leoš Horníček$^{1(\boxtimes)}$ and Zikmund Rakowski^2

^{1} Czech Technical University in Prague, Faculty of Civil Engineering,
Prague, Czech Republic
leos.hornicek@fsv.cvut.cz
^{2} Český Těšín, Czech Republic

Abstract. Tunnel beds are typically formed by massive concrete bodies. In the paper, a mechanically stabilized granular layer with a triaxial geogrid is considered as an alternative solution for tunnel beds. Experimental research was executed on a full scale model of the bed construction in the laboratory. Total settlements, resulting horizontal pressures on the walls and lateral deformations of the base of layers during static and dynamic loading with up to 2 million loading cycles were monitored and evaluated. The tests and results will be described and discussed in the paper. The construction of a mechanically stabilized granular bed derived on this basis can be presented as an innovative and effective solution for infrastructure tunnel projects.

1 Introduction

When slab track is built in railway tunnel, concrete or asphalt bearing layers are usually inserted directly on the tunnel base [3, 11]. This composition ensures stable track geometry and very low plastic deformation of layers. However, necessary elasticity has to be provided by inserting quite expensive elastic elements below the rail or the sleeper, as the concrete or asphalt layers are very stiff and another disadvantage is higher airborne noise emission value [8]. The mass of bed layers is even increasing when tunnel boring machine (TBM) is used. Using this technology results only in circular tunnel profile, which means that bottom part of tunnel have to be backfilled. The overall thickness of bed layers is usually about 1.5 m or even more. Such layers must be homogenous and capable of bearing the imposed loads without significant settlements [4]. From the ecological and economic point of view, it is advisable to consider the possibility of re-utilization of the extracted rock in the bed layers in the case of good quality of the mined rock [1].

In this paper, a mechanically stabilized granular layer (Fig. 1) that e.g. Giroud, Tutumluer, Matys, Belyaev and Rakowski used in their research activities with a stabilizing geogrid is considered as an alternative solution for the construction of tunnel beds. The theoretical assumptions have been subjected to verification under laboratory conditions. Experimental research was executed on a full scale model of the bed construction in the laboratory at the Czech Technical University in Prague. The results of the modelling and its interpretation are described in the paper.

© Springer Nature Switzerland AG 2019
M. Badr and A. Lotfy (Eds.): GeoMEast 2018, SUCI, pp. 169–180, 2019.
https://doi.org/10.1007/978-3-030-01884-9_12

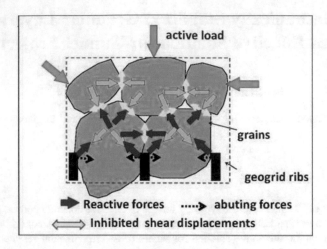

Fig. 1. Interlocking mechanism of a mechanically stabilized layer

2 Concept of Mechanically Stabilized Layer (MSL)

The key mechanism involved in the field of stabilization is the interlocking of aggregates provided by a stiff geogrid structure [2]. During the placement and compaction of a granular layer over the geogrid, the aggregate particles partially penetrate into the apertures and abut against the ribs of the geogrid. This results in the interaction of the geogrid and the aggregates under applied load by locking the aggregate grains in the apertures. In this way, the effect of interlocking is defined as the inhibition of the movement of the particles of a granular layer under applied load [5, 6, 7, 9, 10].

The immobilized stiffened layer is characterized mainly by its thickness, deformability (modulus), both in the in-plane direction and perpendicularly as well, and shear strength. Several researchers e.g. [5, 7] confirmed that a fully immobilized zone can achieve a thickness of 7–10 cm. In the case of a hexagonal geogrid, it can be even up to 15 cm, which means several times the grain size $D_{\max}$, due to firmer abutting of the grains above the geogrid level. Then, logically, the effect of abutting forces is decreasing with the distance from the geogrid, Fig. 2. It means that the layer mechanically stabilized by a stiff geogrid is vertically not homogenous in terms of the mechanical immobilization of grains. Three different zones can be distinguished: the immobilized/stiffened zone h_i, the transition zone h_T, the untouched zone h_o [10]. Obviously, the performance of the whole MSL depends on the intensity of the base zone immobilization, then on the thickness of the zones and the proportion of $h_i{:}h_T$ through the whole MSL construction. The properties of the $h_i{:}h_T$ zones differ for different geogrids and adjacent granular material. According to this concept, MSL has a vertically multilayered structure. In the laboratory model under consideration, there were two geogrid layers, therefore, qualitatively the structure of it is also multilayered as shown in Fig. 3.

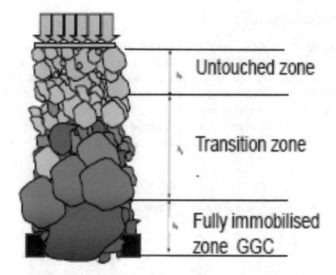

Fig. 2. Layered structure of MSL

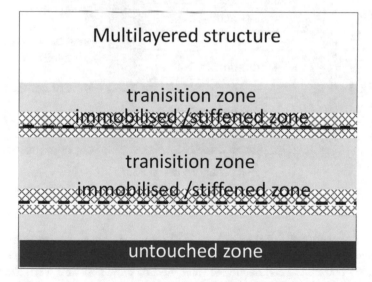

Fig. 3. Structure of a laboratory model

3 Laboratory Model

The above assumptions have been verified through two laboratory models, in which MSL was formed by a different fraction of crushed aggregates and a different type of geogrid. Both models were assembled in the experimental box with dimensions of 2.0 × 1.0 × 0.8 m.

3.1   Description of Model 1

Model 1 (Fig. 4) as assembled as a combination of crushed aggregates graded 0-31.5 and a geogrid with triangular apertures (rib pitch 40 mm). MSL was formed by three layers (250 mm, 250 mm, 150 mm), between which the geogrid was placed. In order to ensure the optimum interlocking of grain aggregates to the mesh of the geogrid, a layer 150 mm in thickness was compacted, further compaction was carried out after the deposition of 100 mm of aggregates, the geogrid and another 40 mm of aggregates. The used aggregates meet the requirements for use in rail constructions in the Czech Republic.

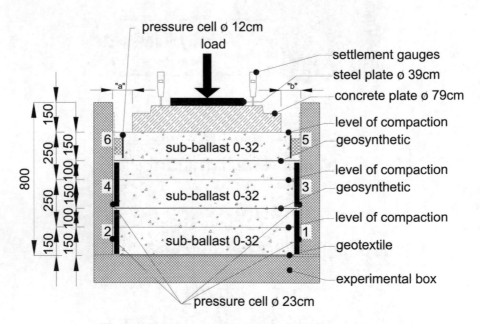

Fig. 4. Model 1 construction composition - cross section

Contact pressure cells of the Geokon 4810 type, which use vibrating wire pressure transducers, 230 mm in diameter, were installed onto both lateral walls (totally four). The Geokon 8002 LC-2x16 datalogger and the LogView software were used for the reading and evaluation of the signals from the contact pressure cells. Both lateral sides were also instrumented with the Glötzl E 12 KF5 type pneumatic pressure cells with a diameter of 120 mm. These pressure cells were situated above the vibrating pressure cells. The Glötzl T1 ALR 16 PF meter was used for reading the signals from these pressure cells. Connecting cables were laid in protective cases.

The levelled aggregate surface was overlaid with a prefabricated circular reinforced concrete cover plate, 140 mm in height and 790 mm in diameter. On this concrete plate, a circular steel plate, 40 mm in height and 390 mm in diameter was placed.

3.2   Measurement Procedure of Model 1

During the model assembly, the settlement of the circular concrete plate was measured at three height levels while a static force of 100 kN (corresponding to overall loading with 0.20 MPa under the reinforced concrete plate) was applied. The load was applied in 4 steps 25 kN in magnitude by the hydraulic actuator with a contact area of 100 mm in diameter. The concrete plate settlement was always read one minute after the respective loading had been reached (after initial consolidation).The measurement of the response on the pressure cells installed on the lateral walls of the experimental box was also carried out in individual loading steps.

After this static loading, the construction was exposed to long-term cyclic loading with a force ranging between 3 kN and 100 kN with a sinusoidal course and a frequency of 3 Hz. The force was applied in the central position of the upper surface of the circular steel plate by means of the loading hydraulic actuator (Fig. 5).

Fig. 5. Measurement of concrete plate settlement and the response of pressure cells

When 1 000, 10 000, 100 000, 200 000, 500 000, 1 000 000, 1 500 000 and 2 000 000 loading cycles were reached, the cyclic loading was interrupted and static loading of the construction with the concrete plate running in the same mode as in the model assembly phase was applied. In individual loading steps, the values on all built-in vibrating and pneumatic pressure cells were read. During cyclic loading, which was applied for a period of 11 days, the surface of granular material was covered with a foil to prevent the construction from undesirable drying.

3.3   Description and Measurement Procedure of Model 2

A combination of crushed aggregates graded 31.5-63 and a geogrid with triangular apertures (rib pitch 40 mm) was used for building up Model 2. This model (Fig. 6, 7) differed from Model 1 not only by using a different aggregate grading (coarse-grained crushed aggregates applied for the ballast bed) and a different type of geogrid, but also by the following parameters:

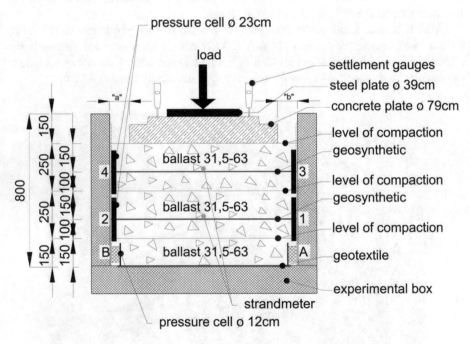

Fig. 6. Model 2 construction composition - cross section

- different position of pressure cells on the lateral walls of the experimental box; by moving pneumatic pressure cells to the bottom a state was reached where the placement level of geogrids corresponded to the centre of vibrating pressure cells,
- mounting 3 vibrating wire strandmeters of the Geokon 4410 type along the longitudinal axis of the geogrids at both height levels to allow a precise measurement of the geogrid deformation under loading (Fig. 8),
- adding a settlement gauge mounted along the longitudinal axis of the experimental box onto the surface of the first geogrid for the measurement of the geogrid vertical movement under loading.

Except for these differences, the assembly phase of Model 2 was the same as in Model 1. In the cyclic loading phase of the construction, the vertical movement of the construction was measured with a built-in settlement gauge after reaching the given number of loading cycles. Like in all the other settlement gauges, the reading was

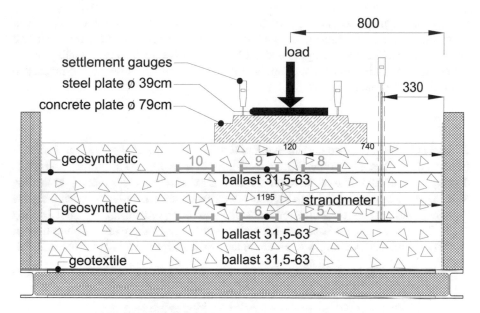

Fig. 7. Model 2 construction composition - longitudinal section

Fig. 8. Strandmeters attached to the longitudinal axis of the geogrid and protected from damage

always taken 1 min after reaching the respective loading state. Besides, the movement
of vibrating wire strandmeters in individual loading states was measured with the
Geokon 8002 LC-2x16 datalogger and the LogView software. The zero reading on the
meters was taken in the construction assembly phase, before covering the respective
layer with aggregates.

3.4   Findings

The data obtained from the measurement of the vertical movement of the concrete plate
(mean values from all 4 settlement gauges) have been processed in the form of charts
displaying settlement in relation to the number of loading cycles for individual loading
states (Fig. 9, 10). In Model 1, a significantly greater settlement of the concrete plate
(which means greater consolidation of MSL) was observed during the first loading
cycle. The value of the subsequent permanent settlement of the plate until the time of
reaching 2 million loading cycles amounted to 1.30 mm, while in Model 2 this

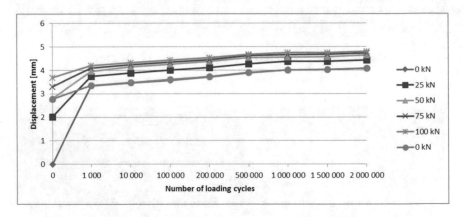

Fig. 9. Settlement pattern during cyclic loading – Model 1

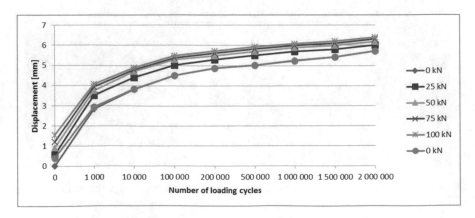

Fig. 10. Settlement pattern during cyclic loading – Model 2

permanent settlement was 5.05 mm. There is also a difference evident from the time pattern and steepness of the settlement curve. A greater settlement in Model 2 is related to using coarse-grained aggregates with a greater volume of voids between the grains and also to the fact that the compaction of the aggregate layer over strandmeters had to be limited for fear of their potential damage. The elastic settlement component (difference between loading with 100 kN and unloaded state) on the contrary, was almost identical in both models.

The values of lateral pressures were obtained after the recalculation of primary data according to the respective calibration curves of individual pressure cells. Due to the fact that a different configuration of pressure cells was used in each of the models, a direct comparison of the measured values cannot be made. The curves displaying the course of values from all pressure cells manifest that no significant changes in the lateral pressure occurred during cyclic loading. Vibrating pressure cells in Model 1 were mounted so that the geogrids would always be situated above or below them, while in Model 2 they were mounted so that the geogrid position would correspond to the horizontal axis of the pressure cells allowing thus the measurement of lateral pressures in the stiffened zone related to the mounting of the respective geogrid.

Those locations are important in the relation to mechanically stabilized layers. In model 1 one pair of gauges was located in not stabilized layer, the second in mechanically stabilized layer. In model 2 lower pair of gauges was on the border between stabilized and not stabilized layers, upper pair in stabilized zone.

The values of the geogrid deformation were obtained after the recalculation of primary data according to the respective calibration curves of individual vibrating wire strandmeters (No. 5–10). The geogrid deformation pattern from all strandmeters under loading with a vertical force of 100 kN is plotted in Fig. 11. The location of individual strandmeters is evident from Fig. 7. The results prove that a slight elongation of the geogrid along its longitudinal axis occurred under the concrete plate due to its settlement. The elongation value under the highest load did not exceed 0.052 mm, which represents a strain of 0.028%.

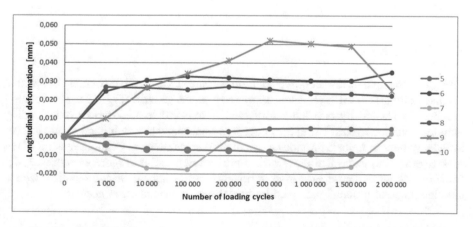

Fig. 11. Geogrid deformation pattern from strandmeters under loading with 100 kN

4 Discussion

The standard solution with massive concrete bed was not a subject of testing as its performance is well known from the literature and practice. The research was aimed on alternative solution with mechanically stabilized layers with stiff hexagonal geogrids. The settlement of the whole structure, side lateral pressures and deformations of geogrids itself were observed.

Additional phase of compaction was visible in both models, in model 1 up to 1000 loading cycles, in model 2 up to 10 000 cycles. The difference was caused by different granularity of the fill, more coarse stone needs more compaction logically. From the functionality point of view, the most important observation is just 1 mm of further settlement up to 2 mil cycles in both models. It is good evidence of the stiffness of the construction, which can be fully acceptable for railway operation.

Side lateral pressure was analysed by the comparison of the performance of different layers. In model 1, the reduction of lateral pressure in stabilized layer confronted with not stabilized varies between 50–75%. In model 2 it is 50% reduction against partly stabilized layer. It forms another evidence of the stiffness of the construction. Strain measurements directly on geogrids showed very low values of strains up to 0, 028% as maximum. The graphs on Fig. 11 confirm the process of additional consolidation of the granular material during first 1–10 ths. cycles. The gauges located under loading plate demonstrated elongation, gauges 7 and 10 located a bit outside the edge of the plate measured compression, as the result of the resistance of mechanically stabilized material. Low values of strain prove low deformability of the system.

The experimental work brought evidences about technological applicability of mechanically stabilized layers as alternative tunnel beds. Deformability of that construction is only marginally greater than classical massive concrete bed. Absolute values of deformations are fully acceptable even for railway tunnels. Obviously, the construction with recycled stone layered with geogrids is costs effective comparing with full massive concrete mixtures. In case of important stone, the cost balance should be considered from case to case depending on local conditions (stone price, transport distance etc.).

5 Conclusions

The main principle of the proposed solution is replacing a massive concrete bed with a mechanically stabilized layer using a properly tested geogrid (Fig. 12). The solution has been evaluated by two laboratory models exposed to long-term cyclic loading ranging 2 million cycles. Using the concept of a mechanically stabilized granular layer in railway beds in tunnels provides a potential of costs savings, possible reusing of the rock (aggregates) excavated during the tunnel driving and it is more environmentally friendly. However, more extensive verification, especially in a combination with a circular shape of a tunnel produced by tunnel boring machines, is needed.

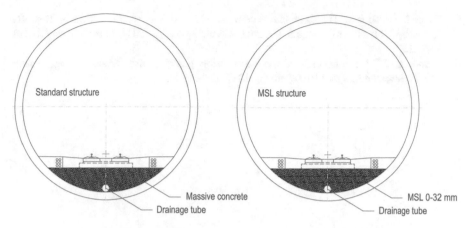

Fig. 12. Proposed new construction of a tunnel bed on the MSL principle

References

1. Berdal, T.: Use of excavated rock material from TBM tunnelling for concrete proportioning. Master thesis, Norwegian University of Science and Technology (2017)
2. Brown, S.F., et al.: Identifying the key parameters that influence geogrid reinforcement of railway ballast. Geotextiles and Geomembr. Elsevier. (2007). https://doi.org/10.1016/j.geotexmem.2007.06.003
3. Darr, E., Fiebig, W.: Feste Fahrbahn. Konstruktion und Bauarten für Eisenbahn und Strassenbahn. Eurailpress Tetzlaff-Hestra, Hamburg (2006)
4. Esveld, C.: Modern railway track, Delft University of Technology (2001)
5. Fischer, S., Szatmari, T.: Investigation of the geogrid-granular soil combination layer with laboratory multi-level shear box test. In: 6th European Geosynthetics Congress, Ljubljana (2016)
6. Jenner, C. et.al.: Trafficking of reinforced, unpaved subbases over a controlled subgrade. In: Proceedings of 7th International Conference on Geosynthetics, Nice (2012)
7. Konietzky, H. et.al.: Use of DEM to model the interlocking effect of geogrids under static and cyclic loading. In: Proceedings of the 2nd International PFC Symposium, Kyoto (2004). https://doi.org/10.1201/b17007-3
8. Lichtberger, B.: Track Compendium. Eurailpress Tetzlaff-Hestra, Hamburg (2005)
9. Matys, M., Baslik, R.: Study of interlocking effect by the push test. In: Proceedings of the 3rd Asian Regional Conference on Geosynthetics GeoAsia2004, Seoul (2004)
10. Rakowski, Z.: An attempt of the synthesis of recent knowledge about mechanisms involved in stabilisation function of geogrids in infrastructure constructions. Procedia Eng. Elsevier. (2017). https://doi.org/10.1016/j.proeng.2017.05.027

11. Tayabji, S., Bilow, D.: Concrete slab track state of the practice. Transp. Res. Rec. J. Transp. Res. Board National Research Council (U.S.) (2001). https://doi.org/10.3141/1742-11
12. Zornberg, J.G.: Stabilization of roadways using geosynthetics. Procedia Eng. Elsevier. (2017). https://doi.org/10.1016/j.proeng.2017.05.048

In-Situ Evaluation of Adhesion Resistance in Anchor Root Zone

Burak Evirgen[✉], Mustafa Tuncan, and Ahmet Tuncan

Anadolu University, Department of Civil Engineering, Eskisehir, Turkey
burakevirgen@anadolu.edu.tr

Abstract. An anchor body is formed by two parts as root zone (bonded length) and the free zone (unbonded length). The main resisting part against the pull out force is the root zone. The similar phenomena is valid for pile or pier foundations, which work according to the side friction arising from external pulling elements. Serious problems can be occur in case the loss of adhesion or frictional resistance between injected grouting material and soil interaction surface. Twelve boreholes were drilled with 22.0 cm in diameter and 2.0 m in length within the aim of assessing this problem. Four different concrete mixtures consist of standard injection mix, gravelly, fine sandy and clayey soils were poured into the holes over 1.0 m from the bottom level. Tremie pipe was also used to elimination of segregation. An anchor cable was placed in the center of each hole with 3.5 m length before the pouring process of mixtures. At the end of 28 days, after completion the setting time of concrete, tendons were subjected to the pull out tests in vertical direction. Pulling force, side friction and displacement values were compared. Thereby, the influence of impurities such as gravel, sand and clay on the adhesion strength were compared that change the mix properties during concrete pouring or injection into the anchor hole.

1 Introduction

Serious irregularities or spaces can occur in the soil structures that already have a rather complex origin, because of environmental factors such as natural formations, stratification, groundwater conditions or tectonic movements. In which case it is necessary to improve the soil properties by using various methods or external support structures. One of them, an anchor is the well-known element that increases capacity by generating extra passive force if the support structures are inadequate in terms of stability. In addition, the root zone created from grouting is the exclusive load carrying part of the anchor systems. While the breaking failure mode is observed in the free zone, on the contrary, the slippage type of failure mode in the grout injected root zone is foreground. Therefore, grouting is a frequently desired requirement in many applications such as soil nails, anchors and rock bolts.

Grouting means the placing of grout material under pressure into caves, joints, fractures and pores of the subsoil to seal or stabilize (Kutzner 1996). Grouting application is used for different purposes for instance fill, compaction or creation of rigid mass to anchoring.

© Springer Nature Switzerland AG 2019
M. Badr and A. Lotfy (Eds.): GeoMEast 2018, SUCI, pp. 181–190, 2019.
https://doi.org/10.1007/978-3-030-01884-9_13

Many researchers have performed experimental or modelling pullout tests in terms of frictional resistance according to numerous variables such as type of soil or grout, property of anchor parts and anchor geometry. Serrano and Olalla (1999) compared the unit bond resistance of short and long rock anchors with failure mechanisms. One of the basic parameters determining the pullout strength of the rock bolts, like the structural element rather than the ground-grout interaction, is the friction behavior between the rock-grout interaction surfaces. Kim et al. (2007) proposed a beam-column finite element model for ground anchors included the interface modeling of soil–grout and grout–strand materials. The load transfer mechanism in the soil–grout interface was simulated as an elasto-perfectly plastic behavior. Field load tests showed that the results on load transfer mechanism were greatly influenced by the modeling of interfaces. Hsu and Chang (2007) evaluated the seventeen full-scale gravity-grouted anchors in a gravel formation in terms of working principle. When the total length of tendon inside the grout or overburden ratio increase, ultimate load increases due to that study. The pullout load per unit bond length was calculated between 143–231 kN/m for shallow tension anchors. Cyr et al. (2013) evaluated the tensile capacity of soil nails, which have metakaolin added grouting material at bonded zone. Because of field pull out tests, it is emphasized that metakaolin reduces both the required cement content and reduces CO_2 emissions. GFRP anti-floating anchors with an interlaminar failure condition were used by Kou et al. (2015) within the scope of prevent the conventional collapse behavior due to sliding between the grouting and weathered soil. The utility of fiber bragg grating sensors that were embedded into the full scaled anchors were verified. Fan et al. (2017) presented the pullout experiments about CFRP anchors located inside the high strength grout via rock mass simulation. It has been indicated that the designed load was increased more than 1000 kN by implementation of two strap ends. Hence, this type of conceptual strengthening elements can be used against grout failure. In case of high-quality grouting, ultimate load value could be reached about 3 times higher than design value in weathered limestone due to experiments with 4 m bonded length realized by Liu et al. (2017). Otherwise, the capacity is lower than expected value caused by insufficient injection. Štefaňák et al. (2017) conducted that seven full-scaled field tests on prestressed ground anchors in neogene clay. The significant difference between ultimate and residual force creates the gradually decreasing of the shear stress along the bond zone.

In the recent popular approaches, some researchers have increased the capacity by assembling of extra members. The innovative tension anchor types such as helical anchors, plate anchors, torpedo anchors and their attached parts exhibit alterations in terms of geometry and working principles (Song et al. 2008, Tian et al. 2014, Tang and Phoon 2016, Wang et al. 2016, Evirgen 2017). The purpose of all approaches is to enhance the frictional capacity by increasing the interaction area between grout and soil.

Despite these studies, the effects of soil materials had a mixing possibility with grouting depending on the disturbance within local conditions have lack of studies. The mixture of grouting material is directly affects both strength against pullout behavior and resistance capability contrary to the environmental influence. The inclusion of natural soil caused from drilling and casting stages of application process, affects the pullout strength due to change of adhesion between grouted body and soil surfaces. On the other hand, aggregates in various dimensions and finer materials can be initially added to the mixture to improve the capacity. Therefore, real-sized field application was carried out within the aim of investigating the effect of grouting properties on the adhesion strength in this paper.

2 In-Situ Application

The real size in-situ application has an essential level of importance within the scope of examination the actual behavior in practical of anchor systems. Therefore, totally twelve vertical anchors were implemented into the clayey type of natural soil which has 12 SPT value, 19.0 kN/m^3 unit weight and 80.0 kN/m^2 cohesion parameters.

2.1 Anchor Details

Each borehole has 2.0 m depth from ground surface. Since the natural clayey soil layers are quite weak under the groundwater level, this depth has been compulsorily selected without encountering water yet. The boreholes were drilled at 2.5 m arrangement as given in Fig. 1 in order to prevent the overlapping of influence zones. 1.0 m bonded length, 1.0 m unbonded length and 1.5 m additional proportion were designed for anchor cable with 3.5 m total length. Seven-coil steel strand with a minimum breaking strength of 265 kN at an elongation of 1.0% and a diameter of 15.2 mm was used as an anchor cable. The modulus of elasticity value was obtained around 200000 MPa. An anchor wedge was used in 46.0 mm length with three segments and the anchor head was used with a single hole.

Fig. 1. Borehole locations

2.2 Grouting Details

Although water and cement suspension are preferred for general use, mixing of gravel, sand and clay particles can be added into the mixture for improving the properties of grouting material according to related standard (TS EN 12715). On the other hand, the content of the mixture may possibly change during the implementation of anchors according to natural soil spillage. In total, four different grouting types were used such as reference mixture and poorly graded gravel (GP), poorly graded sand (SP), low plastic clay (CL) added mixtures separately. Grain size distributions of adding materials are given in Fig. 2. Moreover, grouting details are given in Table 1 with respect to anchors. 0.6 water/cement ratio was used and the additional materials were added to the mixture at 60% level of cement by weight. The rheological properties were chosen to allow the grout to fill the drilled hole without leaving any voids. Besides, pozzolanic cement was used because of the high ratio of fine material content.

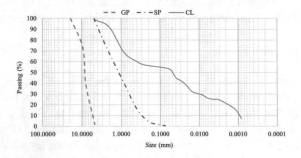

Fig. 2. Additive materials of grouting mixtures, gravel (GP), sand (SP) and clay (CL), respectively

Table 1. Details of grouts

Name of grouting	Additional material to mixture	Anchors
G1	None	A1-A2-A3
G2	Sand	A4-A5-A6
G3	Gravel	A7-A8-A9
G4	Clay	A10-A11-A12

Fig. 3. Modulus of elasticity test setup and specimens at failure state

Three specimens have 10.0 cm in dimeter and 20.0 cm in height, were taken from each mixture to determine the mechanical properties of grouting. After 28 days of curing period, specimens were subjected to the concrete compression test. 0.6 N/mm^2.s pace rate was selected according to TS EN 12390-3. Modulus of elasticity test setup is given in Fig. 3. Within this procedure a 30 t capacity load cell and 50.0 mm capacity four lvdts were used to collect load and displacement values simultaneously. Axial stress and axial strain curves of grouting materials are given in Fig. 4. The clay content decreased the water/cement ratio with increasing the strength of grouting material at the rate of 25%. Due to lack of homogeneous distribution, gravel addition does not provide significant change concerning neither strength nor displacement behavior. On the other hand, sand addition causes a reduction of strength about 15% ratio and increment in 40% level of strain. In the case of poorly graded gravel or sand addition, high degree of water cement ratio prevents the bonds and limits the capacity. The average compressive strength and axial stress values are given in Table 2.

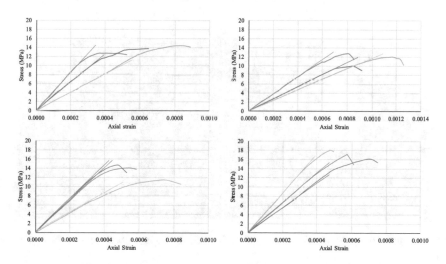

Fig. 4. Stress vs. axial strain graphs of grouting materials

The modulus of elasticity values are calculated concerning the line segment within the proportional limit, which is not exceeding 60% of the peak strength value (FEMA, 2000). These linear sloped lines can be clearly seen in the Fig. 4 for each specimen. Modulus of elasticity of grouting materials were calculated as 31300 MPa, 15000 MPa, 31000 MPa, 34700 MPa for G1 (reference grout), G2 (sand added grout), G3 (gravel added grout) and G4 (clay added grout) types, respectively.

2.3   Application Stages

In this work, anchor application is also started with drilling process as in many geotechnical applications. After placing of steel tendons, required amount of grouting material was poured into the holes. 0.04 m^3 liquid mortar was casted into each hole and

Table 2. Compressive test results of grouts

Name of grouting	Compressive strength (MPa)	Strain (%)
G1	13.6	0.072
G2	11.6	0.101
G3	13.4	0.066
G4	17.1	0.063

finally almost 0.5 m^3 grouting was used in whole application process. A special pro-duced steel tremie pipe was used to prevent segregation during concrete casting. When the boreholes are located at shallow elevation, vibration process is applied in order to provide the most proper placement. Finally, the tendons were centered and grout masses left to take setting for 28 days (Fig. 5).

Fig. 5. Application stages of anchors as a drilling, grouting, vibrating and final position, respectively

2.4   Test Setup

After 28 days to get enough strength of the grout parts, anchors were subjected to in-situ pullout tests according to TS EN 1537. Field test setup consists of following instrumentations; 30 t capacity hydraulic jack, 100 t capacity load cell, 300 mm two displacement transducers (lvdt) for anchor displacement and 100 mm capacity lvdt for ground settlement (Fig. 6). Displacement values were taken from the anchor head elevation during tests to observe the total displacement behavior along borehole included grout movement, tendon shift or elongation.

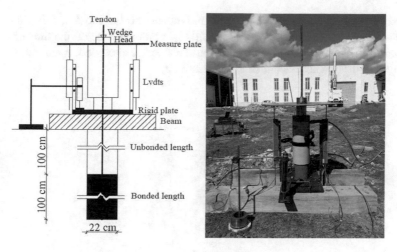

Fig. 6. Field test setup, sketch and photo

3   Results

The pullout force - displacement graphs obtained from the field tests are given in Fig. 7. Each mixture type was evaluated for three different anchors. An average ultimate pullout capacity values for G1 specimens are found as nearly 38.0 kN with 35.0 mm displacement at peak point. If compared with reference anchor, the average anchor capacities increased about 17, 23 and 35% while 18, 17 and 32% displacement decrement in case of sand, gravel and clay addition, respectively. For all cases, the

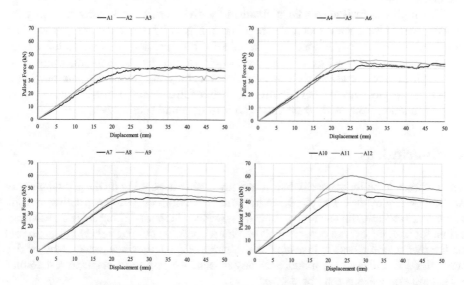

Fig. 7. Pullout force vs. displacement curves of anchors

ground settlement is less than 15.0 mm in average and the most of this value corresponds to the immediate settlement of the fill layer on the surface at the time of loading. As seen in the graphs, after an ultimate pullout load is exceeded, a smooth transition continues without a significant or sharp decrease.

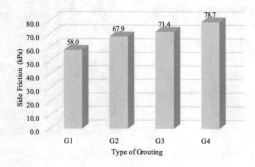

Fig. 8. Ultimate frictional strength of anchors

If pullout capacity of anchors are calculated by using Eq. 1 (Das 1990), nearly 35.0 kN value was achieved. Although this value is approximately equal to the reference injection capacity, it is lower than other anchorage capacities from 15% to 35%.

$$P_u = \pi \, d l c_a \tag{1}$$

Where; P_u is an ultimate resistance of tieback, d is a diameter of bonded zone, l is a length of the bonded zone, c_a is adhesion and it can be taken as 2/3 of undrained cohesion of soil.

The frictional resistance values obtained by dividing the peak stress value by the surface area of the anchor root zone are given in Fig. 8. These values are also valid for unit length of anchor root zone. In spite of axial compression strength decreased, the frictional stress increased by 17% in the anchors formed by the sand-added injection. Although the grout mass starts to moves after reaching the peak value, the maximum frictional capacity caused from adhesive resistance falls, simultaneously. This is known as the transition from static friction behavior to dynamic friction behavior.

4   Conclusions

The increase in pullout capacity becomes unavoidable depending on the growing use of anchorage day by day. Therefore, the frictional behavior between the grouting and soil is crucial. Because, the anchorage capacity how does increased if the friction resistance is insufficient, the overall capacity will be limited. Twelve pullout tests were carried out on the real sized anchors to investigate the change of capacity when gravel, sand or clay was added to the grouting mixture for any reason. According to obtained test results, following conclusions may be drawn.

If the concrete strength used in the grouting increases, the pullout capacity and frictional resistance of anchors increase. However, a homogeneous mixture and good quality placement are required.

The pullout capacity of anchors are increased by 17% to 35% in the case of the addition of mentioned materials to the grout mixture consisting solely of water cement suspension. Similarly, it has been proved that capacity increase can be observed in contrast to the expected case of mixing surrounding soil to the grouting material.

In the field tests, reference grout used anchors have close results to the theoretical values. On the other hand, in the event that fine or coarse material addition, higher pullout capacity values around 25% were obtained from the theoretical calculation. Moreover, field tests were done under relatively fast loading procedure. In case of a residual loading process, the failure behavior will be progressive. Therefore, it is possible that higher capacity values will be obtained at lower displacements.

Acknowledgments. This project was financially supported by the Anadolu University Commission of Scientific Research Projects (Project number: 1705F258). We would like to give our special thanks to the workshop staff of faculty.

References

Cyr, M., Trinh, M., Husson, B., Casaux-Ginestet, G.: Design of eco-efficient grouts intended for soil nailing. Constr. Build. Mater. **41**, 857–867 (2013)

Das, B.M.: Principles of Geotechnical Engineering, 2nd edn. PWS-Kent, Massachusetts, USA (1990)

Evirgen, B.: Evaluation of bearing capacity and sliding potential with new generation applications in scope of geotechnical engineering. Ph.D. thesis, Department of Civil Engineering, Anadolu University, Eskisehir, Turkey (2017)

Fan, H., Vassilopoulos, A.P., Keller, T.: Pull-out behavior of CFRP ground anchors with two-strap ends. Compos. Struct. **160**, 1258–1267 (2017)

FEMA356: Prestandards and commentary for the seismic rehabilitation of buildings. Federal Emergency Management Agency, Virginia, USA (2000)

Hsu, S.-C., Chang, C.-M.: Pullout performance of vertical anchors in gravel formation. Eng. Geol. **90**, 17–29 (2007)

Kim, N.-K., Park, J.-S., Kim, S.K.: Numerical simulation of ground anchors. Comput. Geotech. **34**, 498–507 (2007)

Kou, H.-l., Guo, W., Zhang, M-.y.: Pullout performance of GFRP anti-floating anchor in weathered soil. Tunn. Undergr. Space Technol. **49**, 408–416 (2015)

Kutzner, C.: Grouting of Rock and Soil. A.A. Balkema/Brookfield, Germany, Rotterdam (1996)

Liu, X., Wang, J., Huang, J., Hongwei, J.: Full-scale pullout tests and analyses of ground anchors in rocks under ultimate load conditions. Eng. Geol. **228**, 1–10 (2017)

Serrano, A., Olalla, C.: Tensile resistance of rock anchors. Int. J. Rock Mech. Min. Sci. **36**, 449–474 (1999)

Song, Z., Hu, Y., Randolph, M.F.: Numerical simulation of vertical pullout of plate anchors in clay. J. Geotech. Geoenvironmental Eng. ASCE **134**(6), 866–875 (2008)

Štefaňák, J., Miča, L., Chalmovský, J., Leiter, A., Tichý, P. Full-scale testing of ground anchors in neogene clay. Procedia Eng. **172**, 1129–1136 (2017)

Tang, C., Phoon, K.-K.: Model uncertainty of cylindrical shear method for calculating the uplift capacity of helical anchors in clay. Eng. Geol. **207**, 14–23 (2016)

Tian, Y., Gaudin, C., Cassidy, M.J.: Improving plate anchor design with a keying flap. J. Geotech. Geoenvironmental Eng. ASCE, **140**(5), 04014009, 1–13 (2014)

TS EN 1537: Execution of special geotechnical work- ground anchors. Turkish Standards Institute, Bakanlıklar, Ankara (2001)

TS EN 12390-3: Testing hardened concrete - part 3: compressive strength of test specimens. Turkish Standards Institute, Bakanlıklar, Ankara (2003)

TS EN 12715: Execution of special geotechnical work - grouting. Turkish Standards Institute, Bakanlıklar, Ankara (2003)

Wang, W., Wang, X., Yu, G.: Penetration depth of torpedo anchor in cohesive soil by free fall. Ocean Eng. **116**, 286–294 (2016)

Effect of Yield Criterion on Predicting the Arching Action Around Existing Tunnels Subjected to Surface Loading

Moataz Hesham Soliman[1,2]([✉]), Sherif Adel Akl[2],
and Mohamed Amer[2]

[1] Department of Civil, Environmental, and Construction Engineering,
University of Central Florida, Orlando, FL 32816, USA
moatazhs@knights.ucf.edu
[2] Soil Mechanics and Foundations Laboratory, Department of Public Works,
Cairo University, Cairo, Egypt
saakl@alum.mit.edu

Abstract. Due to the rapid growth of urban development, on many occasions, it may be essential to have construction activities nearby or directly above existing tunnels. In this case, the amount of additional stresses reaching the tunnel becomes a point of concern. The arching action plays the main role in the stress redistributions occurring around the tunnel, and thus the amount of increase in the lining straining actions. Therefore, the goal of this paper is to study how different soil models capture the arching effect on the stress distributions around tunnels. The predicted stresses in any geomechanical problem can be highly dependent on the yield criterion of the material model used in the analysis. The paper compares between the predictions of arching using elastoplastic and poroplastic material models. The Mohr-Coulomb model is used to represent elastoplastic models. Whereas, the Modified Cam-Clay model is used as a poroplastic model. Two-dimensional plane strain models are conducted using the finite element software PLAXIS in which the Mohr-Coulomb and the Modified Cam-Clay are integrated. Although all the used models in this study are calibrated to the behavior of the same soil, different stress redistribution behavior and values are predicted for different constitutive models.

1 Introduction

Due to the complex characteristics of tunneling problems, tunneling engineering is perhaps one of the areas in which numerical methods are more frequently adopted in practice. One important application is to examine the influence of different construction activities, as boring piles (Yao et al. 2008; Dias and Bezuijen 2015) or tunneling (Standing et al. 2015), on existing tunnels nearby. The importance of this application is growing nowadays due to the increasing number of urban tunneling projects leading to existence of large areas of the cities where construction becomes highly restricted. The nature of the stress redistribution influences the load reaching the structure whether it is due to overburden soil, surface surcharge, or lateral earth pressure. These redistributions are highly dependent on the arching action. The arching effect is defined as "the

© Springer Nature Switzerland AG 2019
M. Badr and A. Lotfy (Eds.): GeoMEast 2018, SUCI, pp. 191–198, 2019.
https://doi.org/10.1007/978-3-030-01884-9_14

transfer of pressure from a yielding mass of soil onto adjoining stationary parts" (Terzaghi 1943). Arching may increase or decrease the pressure above a buried structure depending on the soil-structure relative stiffness, known as passive or active arching, respectively (Evans 1983). The use of different material models affects significantly the predicted arching around tunnels and the calculated straining actions in their linings (Soliman 2016).

In this study, a problem of an existing tunnel subjected to a surface load is investigated using different constitutive models. The aim of the study is to compare between the predictions of elastoplastic and poroplastic material models. The Mohr Coulomb (MC) model is used as an elastoplastic model; whereas the Modified Cam Clay (MCC) model is used as a poroplastic model.

The paper first discusses the yielding criterion of each these models. It also shows how to determine the input parameters for each model so that they would represent the same behavior during calibration. Then, the paper illustrates the differences between the predicted stress distributions around the tunnel when different material models are used. The analysis, in this paper, is conducted using the 2D version of PLAXIS V8.2 which has shown a great success in analyzing the behavior of underground structures as tunnels. PLAXIS incorporates both material models used in the study: MC and MCC.

2   Description of Plastic Behavior

An elastoplastic model (MC) and a poroplastic model (MCC) are used to describe the yield criterion in this paper. The yield surfaces of the two models are shown in Figs. 1 and 2. MC yield/failure surface, presented in Fig. 1, is defined in MIT stress space; whereas, the MCC yield surface as well as the Critical State Line (CSL) are presented in Fig. 2. In the elastoplastic model (Fig. 1), the behavior of the material is completely isotropic and elastic for stress state lower than the failure surface. In the poroplastic model, the material exhibits poroplastic behavior when the stress state is inside the ellipsoidal yield surface illustrated in Fig. 2. To match the elastic behavior in the MC model and the poroplastic behavior in the MCC model, the value of Young's modulus should follow Eqs. 1 and 2 (Akl 2015).

$$K = \frac{1+e}{\kappa} * p'$$ (1)

$$E = 3(1 - 2\upsilon)K$$ (2)

where K is the Bulk Modulus, e is the void ratio, κ is the Logarithmic Bulk Modulus, p' is the effective octahedral stress, and v is poisson's ratio. The value of Young's Modulus (E), in any analysis, should be chosen depending on the range of octahedral stress in the same analysis.

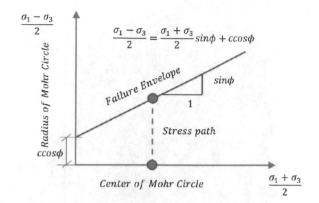

Fig. 1. Mohr-Coulomb failure surface (Akl, 2015)

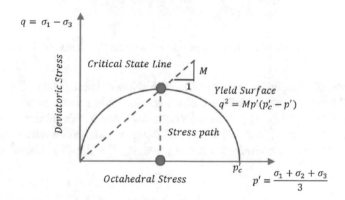

Fig. 2. Modified Cam clay yield surface and critical state line (Akl 2015)

3   Numerical Model

Using the finite element software PLAXIS, a two-dimensional finite element analysis is carried out to model an existing tunnel subjected to surface surcharge load. The analysis follows the plain strain conditions which are typically used in 2D modeling of tunnels. The model configuration is presented in Fig. 3. The tunnel diameter (D) is 8 m and lies at a depth (h) equals to 16 m measured from ground surface to tunnel crown, i.e. h/D = 2.0. The model dimensions are 40 m × 40 m and the tunnel is introduced to its center so that the boundaries have minimum effects on the results. Standard boundary conditions are assigned to the model with vertical and horizontal constraints at the bottom boundary; while the side boundaries are constrained only for horizontal displacements. As shown in Fig. 3, a finite surcharge load is applied at the ground surface above the tunnel. The width of the load is 24 m which is 3 times the tunnel diameter, and it is symmetric around the tunnel centerline. The value of this load is 370 kPa which equals to the overburden pressure above the tunnel springline ($\sigma = \gamma*H = 18.5*20 = 370$ kPa). A homogeneous soil cluster is set in the whole model

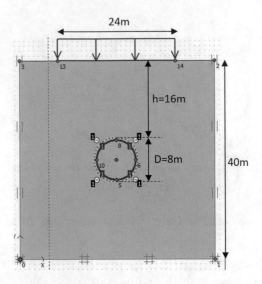

Fig. 3. 2D model of a bored tunnel subjected to surface surcharge in PLAXIS

having material properties representing the North West Sinai Clay (NWSC). The NWSC has the soil properties are illustrated in Table 1 for each of MC and MCC parameters (Odaa, 2017). The value of Young modulus in this table follows the previously discussed Eqs. 1 and 2. The tunnel lining is assigned common values which represents a 40-cm thick reinforced concrete lining: $E = 2.2 \times 10^7$ kN/m^2 and $\rho = 25$ kN/m^3.

Table 1. North-West Sinai clay parameters (Odaa 2017)

Parameter		Value
Soil densities (γ_{unsat}, γ_{sat})		18.5 kN/m^3
Coefficients of permeability (k_x, k_y)		1.86×10^{-5} m/day
Elastic	Young's modulus (E)	3×10^4 kN/m^2
	Poisson's ratio (υ)	0.30
Mohr Coulomb (plastic)	Cohesion (c)	10 kN/m^2
	Angle of internal friction (Φ)	33.5°
	Dilatency angle (ψ)	3.5°
Modified Cam clay (plastic)	Compression index (C_c) or (λ)	0.36
	Recompression index (C_r) or (κ)	0.027
	Void ratio (e)	2.62
	Slope of CSL (M)	1.353

4 Analysis and Results

Figure 4 illustrates a shaded contour display of the deformations around the tunnel using MC and MCC models. These deformation patterns affect the stress distributions in the model as depicted in Fig. 5. In this figure, it is observed that a rotation of the principal stresses takes place around the tunnel. This rotation leads to stress concentrations at the sides of the tunnel (springline) due to tunnel ovalization which acts as a passive state of pressure on the surrounding soil. Whereas, the stresses are arched away from the top of the tunnel, i.e. reduction of stresses at the tunnel crown. This behavior strongly reflects the arching action taking place around the tunnel.

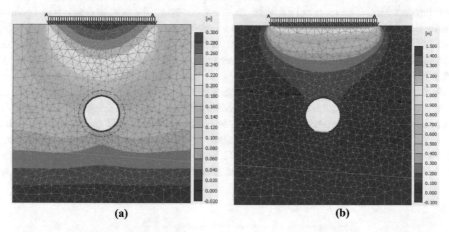

Fig. 4. Deformed mesh after load application using (a) MC Model (b) MCC Model

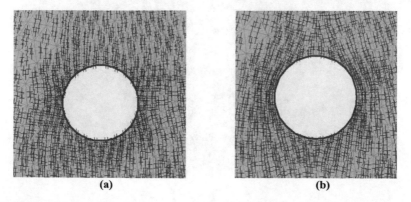

Fig. 5. Principal directions of stresses around a tunnel after surcharge application using (a) Mohr Coulomb and (b) Modified Cam clay material models

Predicted Stress Redistributions

The resulting vertical stresses at each stress point, from each analysis, are subtracted from their corresponding values in case of surface loading without a buried tunnel such that:

$$\Delta\sigma = \sigma_{z(No\ Tunnel)} - \sigma_{z(Tunnel)}$$

The $\Delta\sigma$ values are calculated using MATLAB, and then a contour map is plotted for a study area 32 m × 16 m right above the tunnel and symmetric around its centerline in order to show the stress transfer from the surface load to the tunnel. The 32-m width of the plot is 4 times the tunnel diameter so that it covers more than the 24-m surcharge width. The map height is 16 m, thus, covering the whole depth from ground surface to the tunnel crown. Figures 6 and 7 show the contour maps generated using the MC and MCC analyses, respectively. In both figures, it can be noticed that the changes in the stress are approximately zero ($\Delta\sigma \approx 0$) for the top 8 m of the model (the green area). Thus, the arching effects are negligible in this area. Then, the $\Delta\sigma$ values become relatively significant in the lower 8 m of the studied area above the tunnel. In Fig. 6, positive $\Delta\sigma$ values, up to 35 kPa, are observed right at the top of the tunnel. Therefore, active arching takes place above the tunnel according to Evans (1983). Whereas, negative $\Delta\sigma$ values, around −10 kPa, are noticed on the sides away from the tunnel crown. In case of MCC analysis, Fig. 7 shows significant active arching at the tunnel crown with $\Delta\sigma$ values reaching up to 100 kPa; while the negative

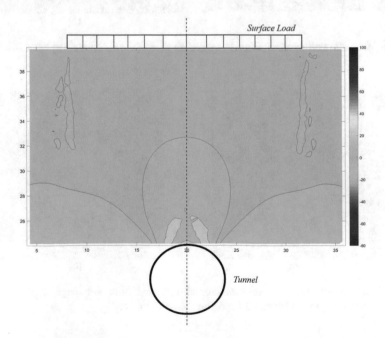

Fig. 6. Distribution of stress changes above the tunnel using Mohr-Coulomb model (kPa)

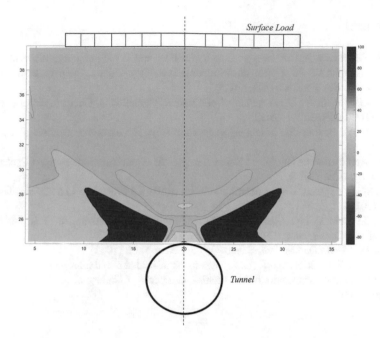

Fig. 7. Distribution of stress changes above the tunnel using Modified Cam clay model (kPa)

$\Delta\sigma$ values on the sides of the crown are up to -80 kPa. Therefore, higher arching effects and stress redistributions are predicted by the MCC model than that when the MC model is used.

5   Conclusion

This paper studied the effect of the yield criterion on the prediction of arching effects and stress redistributions. This was carried out by studying the problem of surface loading above existing tunnels using different elastoplastic and poroplastic constitutive models. Elastoplastic models were represented by the Mohr-Coulomb model (MC); while the Modified Cam Clay was used as an example for the poroplastic models. The study emphasized on the stress transfer from the load at the ground surface to the tunnel considering the stress distributions around the tunnel.

The results have shown that the different models give different predictions of the stress redistributions, however, they are calibrated for the same soil behavior. The arching effects were more significant when MCC model was used as the soil yielding criterion than that when MC model was used. The MCC model predicts higher stress redistributions (active arching) because of the ellipsoidal shape of its yield surface. Stress paths in all directions experience yield in an MCC material. On the other hand, in MC material only near vertical stress paths experience yield while stress paths with low obliquity will continue without inducing any plasticity.

References

Akl, S.A.: Redistribution of stresses due to drilling and depletion using different plasticity models. In: 49th US Rock Mechanics/Geomechanics Symposium. American Rock Mechanics Association, San Francisco, CA (2015)

Dias, T.G., Bezuijen, A.: Data analysis of pile tunnel interaction. J. Geotech. Geoenvironmental Eng. **141**(12), 04015051 (2015)

Evans, C.H.: An examination of arching in granular soils. Massachusetts Institute of Technology (1983)

Odaa, A.: Application of Undrained Stress Analysis to Excavation Side Support System. Cairo University, Giza (2017)

Soliman, M.H.: Arching Around Existing Tunnels Subjected to Surface Loads. Cairo University, Giza (2016)

Standing, J.P., et al.: Investigating the effect of tunnelling on existing tunnels. Underground Design and Construction Conference (2015)

Terzaghi, K.: Theoretical Soil Mechanics. Wiley, New York (1943)

Yao, J., Taylor, R.N., McNamara, A.M.: The effects of loaded bored piles on existing tunnels. In: 6th International Symposium on Geotechnical Aspects of Underground Construction in Soft Ground (2008)

Failure Mechanism of Tunneling in Clay; Solution by GFRP Reinforcement

Zia Hoseini[✉]

School of Civil Engineering, Damghan University,
P.O. Box 36715-364, Damghan, Iran
ziahoseini@std.du.ac.ir

Abstract. There has been an increasing demand for new tunnels in respond to rapid growth in cities. Earth pressure balanced or slurry shield tunneling method is commonly used to construct tunnels in soft ground to improve stability and safety. The use of this kind of reinforcement in tunnel segments allows several advantages mainly related to durability aspects or when provisional lining is forecast. This paper is going to consider the mechanism of failure for tunneling in clay and to find the solution by one of the most respectable kind of reinforcement. The use of GFRP rebars as structural reinforcement in precast tunnel segments, allows several advantages in terms of structural durability or in cases of temporary lining that will have to be demolished later. The application of glass fiber reinforced polymer (GFRP) reinforcement in concrete structures has encountering an increasing interest worldwide, for several applications in civil engineering.

1 Introduction

There has been an increasing demand for new tunnels in respond to rapid growth and needs in urban areas. It is well recognized that tunnel face stability is vital for the safety of tunnel construction in soft ground. The shield tunneling method is commonly used in soft ground to improve stability and safety. Since water and earth pressures at the tunnel face have to be balanced by a supporting medium, either pressurized fluid for slurry shield or spoil of excavated soils for earth pressure balance shield are commonly adopted. No matter what type of shield is used, the design and control of applied pressure at an excavated tunnel face require the pressure to be large enough to maintain face stability (i.e., preventing active failure) but not too large to avoid blow-out at the face (i.e., passive failure). The failure may endanger human life and cause catastrophic damage to the structures within the influence zone. Besides tunnel face stability, soil deformation shall be minimized to ensure a tunnel construction does not inducing distress in adjacent structures and services. The stability of tunnel face has been studied extensively in clay (Broms and Bennermark 1967, Mair 1979, Davis et al. 1980). Mair (1979) provided extensive experimental evidences of the collapse of tunnel face in clay. The extensive experiment evidences can be used to validate the existing analytical solutions based on limit analysis (Leca and Dormeniux 1990, Soubra 2002, Davies et al. 1980). Although many experimental evidences are available for active failure of tunnel face, physical model tests investigating blow-out or passive failure of tunnel face are not common and readily available.

© Springer Nature Switzerland AG 2019

M. Badr and A. Lotfy (Eds.): GeoMEast 2018, SUCI, pp. 199–210, 2019.
https://doi.org/10.1007/978-3-030-01884-9_15

200   Z. Hoseini

Figure 1 shows the profile of undrained shear strength, cu, obtained from the T-bar penetrometer test in Test C2 using "a bar factor" of 10.5 as suggested by Randolph and Houlsby (1984). In the same figure, estimations based on empirical equations proposed by Bolton and Stewart (1994) and Stewart (1992) are also shown for comparison. Besides, the cu profile deduced from vane shear tests is also included. Following Davies and Parry (1982), the cu at 100 g was assumed to be two times the cu obtained using vane shear at 1 g condition.

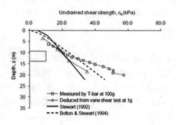

Fig. 1. xxx

For normalized depth z/D < 3.0, the cu profiles obtained from the estimations, T-bar and vane shear measurements are reasonably consistent. For z/D > 3.0, the cu profile obtained from T-bar penetration tests is larger than those from empirical formulas and vane shear tests. Figure 2 shows the cu profile obtained from the T-bar penetrometer test in Test C4. The profiles of cu obtained from vane shear tests as well as empirical formulas are included for comparison. The measured and estimated cu profiles are quite similar.

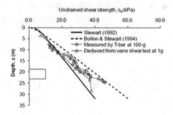

Fig. 2. Undrained shear strength profiles for tunnels located at C/D equal to: (a) 2:1; (b) 4:2 in clay

2   Failure Mechanism

Figure 3 shows the measured normalized displacement vectors on the vertical plane of symmetry at a normalized tunnel face displacement, Sx/D of 0.4 for tunnel located at C/D ratio equal to 2.1. The measured displacement vectors were obtained using the PIV technique and normalized by the tunnel face displacement. The x and z axes are shown in Fig. 5. The measured displacement vectors reveal that the soil in front of the tunnel face is displaced mainly forwards and upwards to the ground surface by the advancing tunnel face. Hence, the soil surface heaves. Some soil in front of the tunnel face is

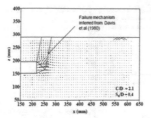

Fig. 3. xxx

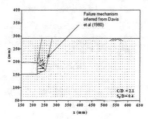

Fig. 4. Normalised displacement vectors for tunnel located at C/D ratios equal to 2.1 in clay: (a) measured; (b) computed

displaced downwards by the advancing tunnel face. A two-block failure mechanism inferred from Davis et al. (1980) is included for comparisons. The failure mechanism consists of two rigid blocks with elliptical cross-sections. The length of the semi-axes perpendicular to the plane of the figure is equal to 0.5D. This is similar to a funnel-type failure mechanism. It is found that the two-block failure mechanism is in good agreement with the observed failure mechanism illustrated by the displacement vectors. This implies that an upper bound solution adopting the failure mechanism may be used to estimate the passive failure pressure. Figure 4 shows the computed normalized displacement vectors at Sx/D of 0.4. The computed displacement vectors are obtained from numerical back-analysis and normalized by tunnel face displacement. The computed displacement vectors also reveal that the soil in front of the tunnel face is displaced mainly forwards and upwards to the ground surface by the advancing tunnel face. It appears that the computed failure mechanism is narrower than the measured one. The computed failure mechanism agrees reasonably well with the two-block failure mechanism by Davis et al. (1980). It is observed that the soil is not displaced downwards significantly by the advancing tunnel face. Figure 6a shows the measured normalized displacement vectors for tunnel located at C/D ratio equal to 4.2. Soil in front of tunnel face is displaced outwards by the advancing tunnel face. The displacement vectors above the tunnel axis are larger than those below it. They illustrate a localized failure mechanism, which differs from the two-block failure mechanism. Figure 6b shows the computed displacement vectors at Sx/D of 0.4. The failure mechanism illustrated by the computed displacement vectors shows fairly close agreement with the localized failure mechanism.

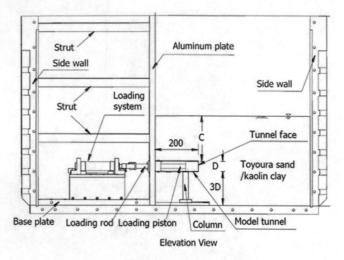

Fig. 5. Basic configuration of a package

3 Passive Failure Pressure of Tunnel Face

Figure 7 shows the variations of the measured tunnel face pressure, σt with Sx/D for tunnels located at C/D ratios equal to 2.1 and 4.2. For tunnel located at C/D ratio equals to 2.1, tunnel face pressure increases with Sx/D but at a reducing rate and reaches a steady state. Lower and upper bound as well as cavity expansion solutions are included for comparisons. The lower bound solution is based on Eq. 1:

$$N = 4\ln\left(\frac{2C}{D} + 1\right) \tag{1}$$

and the upper bound solution is based on the two-block failure mechanism as shown in Fig. 8. Average cu between the ground surface and the tunnel axis is used in the solutions, as cu of the clay layer increases with depth as shown in Fig. 9. The cu value is 12 kPa and the clay layer has a unit weight of 16 kN/m3. The clay level is 1.0 m above the ground surface. It is found that the measured passive failure pressure is closely bounded by the best upper and lower bounds. This is consistent with the good agreement between the two-block failure mechanism and the observed failure mechanism illustrated by displacement vectors in Fig. 8. It should be noted the measured initial tunnel face pressure is 112 kPa, which is slightly lower than hydrostatic tunnel face pressure of 126 kPa at center of the tunnel face. The initial tunnel face pressure at Ko condition is estimated to be 167 kPa assuming a Ko value of 0.6 for lightly over consolidated clay. This is about 55 kPa larger than the measured initial tunnel face pressure. The discrepancy is discussed later. Assuming initial tunnel face pressure is equal to 167 kPa, the measured passive failure agrees well with the upper bound solutions as shown in Fig. 10.

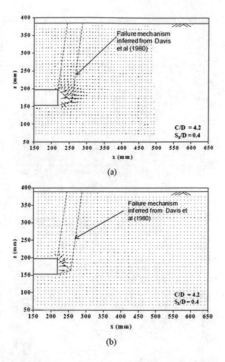

(a)

(b)

Fig. 6. Normalised displacement vectors for tunnel located at C/D ratios equal to 4.2 in clay: (a) measured; (b) computed

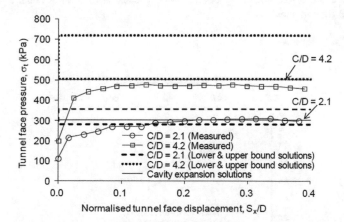

Fig. 7. Variations of measured tunnel face pressure with tunnel face displacement for tunnels located at C/D ratios equal to 2.1 and 4.2 in clay

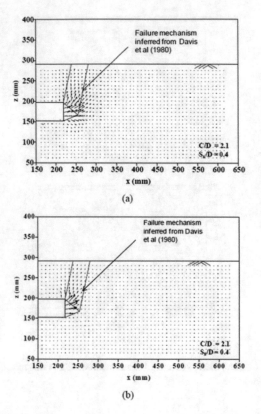

(a)

(b)

Fig. 8. Normalised displacement vectors for tunnel located at C/D ratios equal to 2.1 in clay: (a) measured; (b) computed

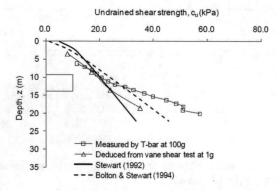

Fig. 9. Undrained shear strength profiles for tunnels located at C/d equal to 2.1 in clay

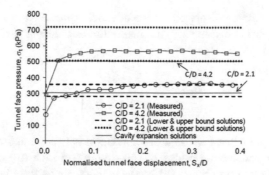

Fig. 10. Variations of corrected tunnel face pressure with tunnel face displacement for tunnels located at C/D ratios equal to 2.1 and 4.2 in clay

4 Surface Heave

The Gaussian distribution (i.e. Equation 2)

$$\Delta = \Delta_{\max} \exp\left(-s^2/2i^2\right) \tag{2}$$

can describe transverse surface settlement induced by single tunnel construction. Mair and Taylor (1997) found that the value of K varies from 0.4 to 0.6 for clay. In this study, a two-dimensional Gaussian distribution is adopted and defined as Eq. 2. Where (x_0, y_0) is the location of the maximum surface settlement. Instead of fitting settlement profiles, the Gaussian distribution is fitted to the measured heave values as a first approximation. It should be noted surface heave due to tunnel face displacement may not be symmetrical. Volume under the two-dimensional Gaussian distribution, V_s, is obtained from the integration of Eq. 3.

$$\Delta = \Delta_{\max} \exp\left(-\left[\frac{(x - x_0)^2 + (y - y_0)^2}{2i^2}\right]\right) \tag{3}$$

Since the volume under the Gaussion distribution, V_s is given by Eq. 4.

$$V_s = 2\pi\Delta_{\max}i_x i_y \tag{4}$$

Substituting $i = i_x = i_y$, (Eq. 5)

$$V_s = 2\pi\Delta_{\max}i^2 \tag{5}$$

the volume displaced by the advancing tunnel face is given by Eq. 6.

$$V_t = \frac{\pi D^2}{4} S_x \tag{6}$$

Assuming the volume displaced by the advancing tunnel face is equal to the volume under the Gaussian distribution, the maximum normalized surface heave is given as follows: (Eq. 7).

$$\Delta_{max} = \frac{S_x}{8\left(K\left[\frac{C}{D}+0.5\right]\right)^2} \tag{7}$$

Equation 7 implies that there is a linear relation between maximum surface heave and tunnel face displacement. Besides, the maximum surface heave decreases with C/D ratio. Substituting Eq. 7 into Eq. 3, surface heave induced by the advancing tunnel face can be calculated as Eq. 8.

$$\Delta = \frac{S_x}{8\left(K\left[\frac{C}{D}+0.5\right]\right)^2}\exp\left(-\left[\frac{(x-x_0)^2+(y-y_0)^2}{8\left(KD\left[\frac{C}{D}+0.5\right]\right)^2}\right]\right) \tag{8}$$

The results of centrifuge model tests and finite element back-analyses investigating the passive failure and deformation mechanism due to tunneling in sand and clay are reported. Soil failure mechanisms and passive failure pressures of tunnel face are described and discussed. In addition, the induced ground surface displacements due to the increase in tunnel face displacement are also reported. In clay, excess pore water pressure due to an increase in tunnel face displacement is reported. Besides, long-term settlement and dissipation of excess pore water pressure after passive failure are also discussed. For tunnel located at C/D ratio equal to 4.2 in clay, the displacement vectors reveal a localized failure mechanism. The measured passive failure pressure is close to the lower bound and cavity expansion solutions. This may be due to the similarity between the spherical cavity expansion and the localized failure mechanism. However, there is a large discrepancy between the measured failure pressure and the upper bound solution attributed to the large discrepancy between the localized failure mechanism and the two-block failure mechanism adopted in the upper bound solution. The computed passive failure is larger than the measured value. The longitudinal and transverse surface heave may be described by the two dimensional Gaussian distributions.

5 GFRP Reinforcement Solution

The use of Glass Fiber Reinforced Polymer (GFRP) rebars in concrete structures is an innovative solution that can be proposed in alternative to the traditional steel rebars, mainly when a high resistance to the environmental clay's attack is required. Indeed, in comparison with steel, GFRP does not suffer corrosion problems, presents higher tensile capacity, and lower weight (Nanni 1993; Benmokrane et al. 1995; Alsayed et al. 2000). The material is also non-conductive for electricity and non-magnetic. Nevertheless, it has to remark that this type of reinforcement is not suitable for all the applications. Firstly, the cost of the GFRP rebars is generally higher with respect to the traditional steel ones, moreover the GFRP reinforcement is affected by problems related to static fatigue when subjected to high-level long-term tensile stresses (Almusallam et al. 2006). Furthermore, the structural effects of the low value of the Young's modulus and the bond behavior (Cosenza et al. 1997; Yoo et al. 2015) have to be considered. Finally, the durability performance of GFRP bars could be improved by selecting the appropriate constituents of the composites (Micelli and Nanni 2004; Chen et al. 2007). The possibility of using a GFRP reinforcement in precast tunnel segments is investigated herein. In mechanical excavated tunnels, generally made with a Tunnel Boring Machine (TBM), the lining is composed of precast elements, placed by the TBM during the excavation process and used as reaction elements during the advancing phase. The adoption of precast segments reinforced with GFRP rebars is suitable when durability problems could jeopardize the tunnel integrity (e.g. aggressive environmental as in clay tunnels or presence of aggressive soils). Furthermore, the possibility of using non-metallic reinforcement allows a strong reduction of the concrete cover, avoiding problems of crushing during the segments handling. The GFRP reinforcement is also suitable in parts of the tunnel that have to be eventually demolished. Typical examples are metro or railways lines, when the station is built after the tunnel excavation or in road tunnels when the section of the tunnel has to be modified for creating safety areas. Finally, the use of this technology is adequate for creating dielectric joint in tunnels. The adoption of GFRP rebars in substitution to traditional steel cage can be proposed in a series of tunnels affected by the afore-mentioned problems. The higher cost of the material, with respect to traditional steel cages, can be balanced if the overall costs related to the tunnel construction and maintenance is considered. Furthermore, the lining at final stage is often mainly in compression and the static fatigue in the GFRP rebars is limited. On the structural point of view it is important to demonstrate to the designers that the behavior of precast tunnel segments reinforced with GFRP rebars is comparable (or even better) with respect to the traditional steel reinforcement. With this purpose, the simulated model on precast segments subjected to bending actions have been performed in order to compare the structural behavior of elements reinforced with GFRP and traditional steel bars. The segments were subjected to bending actions in order to highlight the effect of the reinforcement. Bending simulation is also representative of the provisional loading stages as demolding, storage and handling.

The reference segment has been cast with a traditional steel cage made of 12Ø12 bars placed in the intrados and 12Ø12 bars at the extrados surfaces (Fig. 11a). The steel rebars are classified, according to EC2 (EN-1992-1, 2008), as B500 with characteristic yielding stress (fyk) equal to 500 MPa. The GFR segment is reinforced with GFRP bars with a characteristic tensile strength (ftk) equal to 690 MPa. In particular, the reinforcement consists in 12Ø12 longitudinal bars at the extrados surface and 12Ø14 longitudinal bars at the intrados surface (Fig. 11b). The GFRP reinforcement has been designed, according to the codes indications (Fib Bulletin 40, 2007; CNR-DT203, 2007), in order to provide the same ultimate bending resistance of the reference SR segment. The characteristic and design properties of the materials, used for the evaluation of the ultimate bending moments of GFR and SR elements, are summarized in Table 1. With reference to the concrete in compression, the safety coefficient (cc) is assumed equal to 1.5, according to (EN-1992-1, 2008); the long term behavior factor ac is set equal to 1 (EN- 1992-1, 2008). For what the reinforcement is concerned, it is

Table 1. Design values of the material properties for the M-N envelopes definition.

Material	Characteristic values	Safety coefficients	Design values
Concrete C40/50	$f_{ck} = 40$ MPa	$\gamma_c = 1.5$; $_c = 1$	$f_{cd} = \frac{f_{ck} \cdot \alpha_c}{\gamma_c} = 26.7$ MPa
Steel	$f_{yk} = 500$ M Pa	$\gamma_s = 1.15$	$f_{yd} = \frac{f_{yk}}{\gamma_s} = 435$ MPa
CFRP	$f_{tk} = 690$ MPa $\varepsilon_{fk} = \frac{f_{tk}}{E_f} = 1.64\%$	$\eta_a = 0.7$; $\eta_l = 1$; $\gamma_f = 1.5$ $\alpha = 0.9$	$f_{td} = \frac{f_{tk} \cdot \eta_a \cdot \eta_l}{\gamma_f} = 322$ MPa $\varepsilon_{fu} = \alpha \frac{\varepsilon_{fk} \cdot \eta_a}{\gamma_f} = 0.69\%$

$^a E_f = 42$ GPa (mean value).

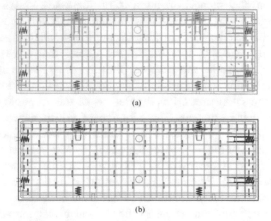

(a)

(b)

Fig. 11. Detailing for the traditional reinforcement (a) and the GFRP reinforcement (b)

worth remarking the different value of the safety coefficient, adopted in the design, equal to 1.15 (cs) and 1.5 (cf) for the steel reinforcement and for GFRP rebars, respectively. Furthermore, in order to account for the possible degradation of the GFRP mechanical properties under specific environmental conditions, such as exposition to moisture, typical of tunnel linings, the environmental conversion factor (ga) is applied, ad set equal to 0.7 according to (CNRDT203, 207). The conversion factors for long-term effects (gl), accounting for possible degradation of the GFRP mechanical properties due to creep, relaxation, and fatigue, is set equal to 1, as suggested in (CNR-DT203, 2007), for Ultimate Limit State analyses.

6   Conclusions

GFRP rebars can be a solution in some problems that can arise in the segmental lining construction. The suitability of this technique was investigated with full-scale tests and the obtained results allow drawing the following remarks:

– On the point of view of the flexural structural behavior, there are not significant differences when the steel reinforcement is substituted with a GFRP reinforcement.

In fact, despite the brittleness of the material, the performed flexural full-scale test showed that the structure exhibited not only a significant strength but also an adequate ductility;

- The segments reinforced with GFRP bars tested under TBM thrust loads exhibited a suitable behavior. The maximum crack widths, even under exceptional loads, were always lower than the allowable ones given by the codes;
- The design approach suggested by the codes appears in accordance with the tests evidences. Nevertheless, the adoption, for the GFRP, of safety coefficients remarkably on the safe side, can lead to a structural over-strength, that could be penalizing, mainly for temporary structures.

Finally, the use of GFRP reinforcement appears a very interesting and promising solution in tunnel segmental lining, in some critical situations that often are present in tunnel construction. The choice of adopting GFRP reinforcement instead of a typical steel one, can be particularly advantageous when durability problems are foreseen, for provisional elements to be eventually demolished, and when the necessity of creating dielectric joint in tunnels arises.

References

Alsayed, S.H., Al-Salloum, Y.A., Almusallam, T.H., 2000. Performance of glass fiber reinforced plastic bars as a reinforcing material for concrete structures. Composite: Part B 31, 555–567

Benmokrane, B., Chaallal, O., Masmoudi, R.: Glass fibre reinforced plastic (GFRP) rebars for concrete structure. Constr. Build. Mater. **9**(6), 353–364 (1995)

Bolton, M.D., Stewart, D.I.: The effect on propped diaphragm walls of rising groundwater in stiff clay. Geotechnique **44**(1), 111–127 (1994)

Broms, B.B., Bennermark, H.: Stability of clay at vertical openings. J. Soil Mech. Fdns Div. Am. Soc. Civ. Eng. **193** SM1, 71–94 (1967)

Chen, Y., Davalos, J.F., Kim, H.-Y.: Accelerated aging tests for evaluations of durability performance of FRP reinforcing bars for concrete structures. Compos. Struct. **78**, 101–111 (2007)

Cosenza, E., Manfredi, G., Realfonzo, R.: Behavior and modeling of bond of FRP rebars to concrete. J. Compos. Constr. ASCE **1**(2), 40–51 (1997)

Davis, E.H., Gunn, M.J., Mair, R.J., Seneviratne, H.N.: The stability of shallow Tunnels and underground openings in cohesive material. Geotechnique **30**(4), 397–416 (1980)

Davies, M.C.R., Parry, R.H.G.: Determining the shear strength of clay cakes in the centrifuge using a vane. Geotechnique **32**(1), 59–62 (1982)

Leca, E., Dormieux, L.: Upper and lower bound solutions for the face stability of shallow circular tunnels in frictional material. Geotechnique **40**(4), 581–606 (1990)

Mair, R.J., Taylor, R.N.: Bored tunneling in the urban environment. Proc. 14th Int. Conf. Soil Mech. Found. Eng. **4**, 2353–2385 (1997)

Nanni, A. (Ed.): Fiber-Reinforced-Plastic (GFRP) Reinforcement for Concrete Structures: Properties and applications. Developments in Civil Engineering, vol. 42, p. 450. Elsevier Science (1993)

Micelli, F., Nanni, A.: Durability of FRP rods for concrete structures. Constr. Build. Mater. **18**, 491–503 (2004)

Randolph, M.F., Houlsby, G.T.: Limiting pressure on a circular pile loaded laterally in cohesive soil. Geotechnique **34**(4), 613–623 (1984)

Soubra, A.H.: Kinematical approach to the face stability analysis of shallow circular tunnels. In: Proceedings of the Eight International Symposium on Plasticity, pp. 443–445 (2002)

Yoo, D.-Y., Kwon, K.-Y., Park, J.-J., Yoon, Y.-S.: Local bond-slip response of GFRP rebar in ultra-high-performance fiber-reinforced concrete. Compos. Struct. **120**, 53–64 (2015)

Author Index

A
Akl, Sherif Adel, 191
Alberto-Hernandez, Yolanda, 159
Albuquerque, Paulo J. R., 31
Amer, Mohamed, 191

B
Bahi, Lahcen, 53
Beghoul, Mohammed, 42
Bhatt, Devendra Datt, 19

C
Charif, Khalil, 19
Chikhaoui, Mohamed, 71

D
Demagh, Rafik, 42

E
Evirgen, Burak, 181

G
Garcia, Jean R., 31

H
Hemeda, Sayed, 126
Horníček, Leoš, 169
Hoseini, Zia, 199

M
Meguid, Mohamed A., 81
Mucheti, Alexsander Silva, 31

N
Naouz, Yassir, 53
Nechnech, Ammar, 71

O
Ouadif, Latifa, 53

P
Perminov, Nicolay, 94

R
Rakowski, Zikmund, 169
Rao, K. S., 105
Rustum, Rabee, 19

S
Sharif, Emad Y., 1
Sidhu, Hardev, 1
Singh, L. M., 105
Singh, T., 105
Soliman, Moataz Hesham, 191

T
Tuncan, Ahmet, 181
Tuncan, Mustafa, 181

© Springer Nature Switzerland AG 2019
M. Badr and A. Lotfy (Eds.): GeoMEast 2018, SUCI, p. 211, 2019.
https://doi.org/10.1007/978-3-030-01884-9

Printed in the United States
By Bookmasters